HOW TO DESIGN ELECTRONIC PROJECTS

ALSO BY THE SAME AUTHOR

No. 222	Solid State Short Wave Receivers for Beginners
No. 223	50 Projects Using IC CA3130
No. 224	50 CMOS IC Projects
No. 226	How to Build Advanced Short Wave Receivers
No. 227	Beginners Guide to Building Electronic Projects
No. BP45	Projects in Opto-electronics
No. BP49	Popular Electronic Projects
No. BP56	Electronic Security Devices
No. BP59	Second Book of CMOS IC Projects
No. BP65	Single IC Projects
No. BP69	Electronic Games
No. BP71	Electronic Household Projects
No. BP74	Electronic Music Projects
No. BP76	Power Supply Projects
No. BP80	Popular Electronic Circuits – Book 1
No. BP83	VMOS Projects
No. BP91	An Introduction to Radio DXing
No. BP94	Electronic Projects for Cars and Boats
No. BP95	Model Railway Projects
No. BP96	CB Projects
No. BP98	Popular Electronic Circuits – Book 2
No. BP99	Mini-matrix Board Projects
No. BP103	Multi-circuit Board Projects
No. BP105	Aerial Projects
No. BP106	Modern Op-Amp Projects
No. BP107	30 Solderless Breadboard Projects – Book 1
No. BP110	How to Get Your Electronic Projects Working
No. BP113	30 Solderless Breadboard Projects – Book 2
No. BP117	Practical Electronic Building Blocks – Book 1
No. BP118	Practical Electronic Building Blocks – Book 2
No. BP121	How to Design and Make Your Own P.C.B.s
No. BP122	Audio Amplifier Construction
No. BP123	A Practical Introduction to Microprocessors
No. BP130	Micro Interfacing Circuits – Book 1
No. BP131	Micro Interfacing Circuits – Book 2

HOW TO DESIGN ELECTRONIC PROJECTS

by
R. A. PENFOLD

BERNARD BABANI (publishing) LTD
THE GRAMPIANS
SHEPHERDS BUSH ROAD
LONDON W6 7NF
ENGLAND

PLEASE NOTE

Although every care has been taken with the production of this book to ensure that any projects, designs, modifications and/or programs etc. contained herein, operate in a correct and safe manner and also that any components specified are normally available in Great Britain, the Publishers do not accept responsibility in any way for the failure, including fault in design, of any project, design, modification or program to work correctly or to cause damage to any other equipment that it may be connected to or used in conjunction with, or in respect of any other damage or injury that may be so caused, nor do the Publishers accept responsibility in any way for the failure to obtain specified components.

Notice is also given that if equipment that is still under warranty is modified in any way or used or connected with home-built equipment then that warranty may be void.

First Published – March 1984

British Library Cataloguing in Publication Data
Penfold, R. A.
How to design electronic projects. – (BP127)
1. Electronic apparatus and appliances – Amateurs' manuals
I. Title
621.381 TK9965

ISBN 0 85934 102 X

Printed and bound in Great Britain by Cox & Wyman Ltd, Reading

PREFACE

Most electronic project constructors sooner or later move away from making exact copies of published designs and begin to modify projects to suit their own requirements. It is then only a small step, from adapting parts of published circuits to fit a custom made project, to designing ones own equipment from scratch. It is in fact possible to put together ones own designs with very little technical knowledge, although this inevitably leads to rather hit and miss techniques which are rather slow and can lead to costly mistakes. Once a basic understanding of electronic components and circuits has been gained it is not too difficult to apply a more reasoned approach to electronic project design, and projects to suit ones precise needs can then be designed more rapidly and with greater likelyhood of success.

There is a lot of information available on various circuit building blocks in catalogues, books, magazines, etc., but there is little information that helps the amateur user to integrate building blocks into practical projects. The aim of this book is to help the reader to put together projects from standard circuit blocks with a minimum of trial and error, but without resorting to any advanced mathematics. Hints on designing circuit blocks to meet your special requirements where no "stock" designs are available are also provided. The subject is tackled by taking a series of simple practical examples, analysing exactly what each circuit must do, exploring possible methods of achieving each circuit action, and then working out practical designs including all circuit values. Thus a number of useful circuits are provided in addition to project design advice.

R. A. Penfold

CONTENTS

Getting Started

When first experimenting with project designs it is very easy to get carried away with ambitious projects that ultimately come to nothing due to a lack of the necessary expertise. A lot of useful experience may be obtained even if the end result is failure, but it is probably better to initially confine ones activities to fairly simple projects where the chances of a successful outcome are far greater, and the likelyhood of getting frustrated and giving up electronics altogether are much less!

It is important to have as much background information as possible, and the widest possible knowledge of electronics components and circuits. There are plenty of books and articles which give basic electronics theory and it is advisable to study some of these thoroughly. As important points of theory crop up they will be discussed in this book, but there is simply not enough space here to give a complete course on electronics theory.

Having gained a reasonable understanding of basic electronic principles a useful next step is to study some practical designs to see how these principles are put to use in practice. Most published designs are accompanied by some sort of circuit description, and many are complete with a very detailed explanation of the circuit. Select a few designs that are not over complex and see if you can understand the function of every component in each design. If the circuit description provided is only very brief try to work out the function of each component for yourself. This can be very instructive, but it is best to choose circuits based on discrete semiconductor devices or common ICs such as operational amplifiers since the function of discrete components in circuits which use specialised ICs is often difficult or impossible to determine simply by looking at the circuit diagram. A capacitor wired from one pin of a complex and specialised IC to (say) the negative supply rail could be a timing component in a monostable or astable circuit, a decoupling capacitor, a frequency compensation component, a filter capacitor, or any of numerous other possibilities.

Apart from looking at designs component by component it is a good idea to look at the overal make-up of the design, and the various stages that fit together to give the desired function. There may be a block diagram which explains the general make-up of the design, but if there is not you should be able to produce such a diagram yourself. Selecting suitable circuit blocks and fitting them together to produce a complete design giving the required function is really the primary topic of this publication, but it is advisable to study as many designs as possible in addition to those provided here as a lot of useful ideas can be gained in that way.

If you study a substantial number of electronic circuit designs you may notice that some types of circuit block appear in two different designs from two different designers. In fact you may find circuit blocks that appear a great many times in a variety of projects from different sources. The reason for this is simply that no one normally sits down and works out an electronic design entirely from scratch. There are many well known and much used circuit blocks which can be used in designs, and it makes sense to use these where possible rather than to waste time designing a building block which will work no better than one based on a well known configuration. The well known circuit configurations are well known and popular because they represent what is probably the best way of providing a particular function, and using an alternative will usually just give added complexity or inferior results.

You could gain from this the impression that putting together an electronic design is just a matter of finding the right electronic circuit blocks and fitting them together in the appropriate way. This may sometimes be the case, but in most applications there will be parts of the circuit where no common circuit block provides the required function, and it is then up to the designer to produce a circuit (or modify an existing configuration) to give the required circuit function. It is here that a creative designer can use his or her imagination to produce a good working design. Also, interfacing circuit blocks to one another is not just a matter of connecting the output of one block to the input of another, as it will often be necessary to make modifications to one or both building blocks before they will fit together satisfactorily, and in many cases it is necessary to add a small circuit to interface the two circuit blocks properly.

Another point worth bearing in mind is that the various circuit blocks are nothing and of no value in themselves, and it is up to the designer to see the possibilities and to make the circuit blocks perform a useful task. This pairing of project ideas on the one hand with the appropriate electronic building blocks on the other may seem to be the most simple part of project design, but it is one of those things where it seems obvious once the correct circuit blocks have been selected and placed into an overal design in the correct manner, but can seem to be impossible at times when trying to work out a design. If you can manage to successfully negotiate this initial stage of the design procedure it is likely that you will be able to overcome the other obstacles and eventually produce a working design.

It is obviously an advantage to have a wide range of electronic circuit blocks available for use in your designs, and there are many possible sources for these. It is also essential to have at least basic data on a reasonable range of semiconductor devices including transistors, thyristors, diodes, and popular ICs such as operational amplifiers. Manufacturers data books are perhaps the best source for both types of information, although many of these books may be difficult for the amateur user to obtain. Some of the larger component catalogues produced by electronic component mail order companies provide a lot of useful data which is not normally very extensive for any given device, but is often sufficient to enable a suitable device to be chosen. There are many books which contain useful circuits (such as books number BP117 *Practical Electronic Buliding Blocks – Book 1*, and number BP118 *Practical Electronic Building Blocks – Book 2*, by the same publisher and author as this publication) also magazine articles giving helpful circuit information appear quite frequently. Apart from information which is provided specifically to aid the circuit designer, it is possible to "borrow" sections of any published circuit for your own use. Any publication that provides circuits can therefore be a useful source of information.

Simple Timer

As our first example of a simple practical project we will consider a simple timer circuit which must give a two tone audible alarm of some kind at twelve switched intervals from half a minute to six minutes after the unit has been switched on. In other words a typical timer circuit of the "Egg Timer" variety that appear from time to time in technical publications.

It is unusual for there to be just one possible way of tackling an electronic project, and there is often a variety of ways in which the desired function can be obtained. One of the reasons for selecting a simple timer as the first demonstration project is that there are so many possible ways of achieving the desired effect. There are at least three integrated circuits which are ideal for this type of timing application, and there are several semiconductor devices which can easily be used in timing circuits. The same is true of the audio alarm generator part of the unit, with an almost unlimited number of possible ways of generating a simple two tone sound of some kind.

Whatever circuit blocks are used, the general arrangement of the circuit will be something along the lines suggested in the block diagram of Figure 1. Here we have the timer circuit which must be triggered automatically when the unit is switched on, and must have the desired twelve switched delay times. This must switch on the audio alarm generator at the end of the timing period, but note that two stages are needed to generate the alarm signal. This is simply due to the lack of an audio oscillator which has built-in frequency modulation, and it is therefore necessary to have one oscillator to generate the basic audio alarm signal, and a second (low frequency) oscillator to vary the frequency of the audio oscillator to give the two tone effect.

If we only consider the timer section for the time being, in order to select a suitable type of circuit it is first necessary to consider any essential features of this circuit. The two obvious ones are that it must be possible to have the timing run start automatically at switch-on, and the circuit must be capable of giving the required delay times. In fact these criteria do not really narrow down the range of possibilities very much since

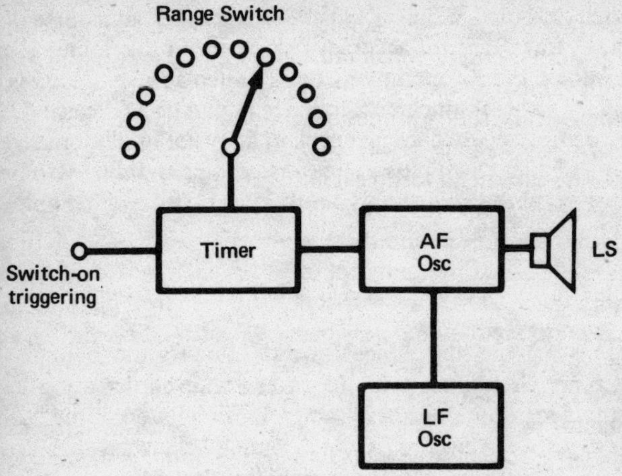

Fig. 1. A suitable arrangement for the simple timer unit

most simple timer circuits are capable of auto-triggering at switch-on (and this feature could be easily added to any that do not have an innate auto-triggering facility), and most simple timer circuits are just about capable of giving the range of times that are needed.

Two important points which must be considered are whether the circuit will be battery or mains powered, and what degree of accuracy is required. There are timer circuits which operate by using the 50 Hertz mains signal to provide a timebase, but this type of circuit is obviously not applicable in this case if battery power is to be used. Timers of the type under consideration here are normally small self contained units, and therefore battery powered. We will assume that a battery powered unit is required in this instance.

A very high degree of accuracy is not really of great

importance as the timer is not automatically operating some piece of equipment, but is simply giving an audible signal which indicates to someone that they should take a certain course of action. A low degree of accuracy is inherent in this round-about way of doing things, and an error of as much as a few seconds is unlikely to be of much consequence in practice. Precision timer circuits using techniques such as a crystal oscillator stage driving a divider chain plus some control logic is unlikely to be worthwhile, and would simply result in a grossly over complex and expensive finished unit.

C – R Timer

This really narrows the choice down to some form of simple C – R timer circuit, but this still leaves a wide choice of options since there are numerous devices which can be used as the basis of this type of timing circuit. Semiconductor devices such as JFETs, VMOS transistors, and unijunction transistors can all be used in this type of circuit, but these days it is probably better to use an integrated circuit as the basis of the design as this is likely to give more predictable results and better reliability. The obvious choice for this application is the 555 timer IC, and operational amplifiers are very versatile and definitely worthy of consideration.

In this application the 555 must be used in the monostable mode where a single output pulse is provided when the device is triggered, rather than in the astable (oscillator) mode. The 555 is a well known and much used device, and the basic monostable configuration of Figure 2 will probably be familiar to many readers.

In order to trigger the circuit pin 2 of IC1 must be briefly pulsed low, and in this case we require the circuit to trigger at switch-on. This is achieved using R1 and C1 which provide a brief negative pulse when power is initially applied to the circuit. C1 is uncharged at first and therefore takes pin 2 to the negative supply potential, but it soon charges to the full supply potential so that pin 2 is taken above the one third of the supply potential trigger threshold. With some types of monostable the trigger input could simply be taken direct to the negative supply

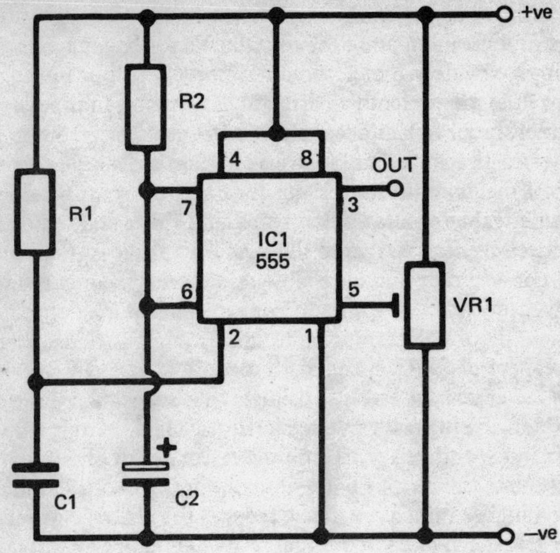

Fig. 2. The basic 555 monostable configuration

rail, but this method of triggering cannot be used with the 555 because the output pulse cannot end until the trigger input is taken above the trigger threshold voltage.

The output of the 555 is pin 3, and this is normally low but goes high when the device is triggered. The length of the positive output pulse is determined by the values of timing components R2 and C2, and the output pulse ends when the voltage on C2 reaches about two thirds of the supply voltage. This takes about 1.1CR seconds, and for this type of calculation it is more convenient to use microfarads and megohms than farads and ohms.

In an application of this type where fairly long pulse lengths are needed a simple C – R timer is barely able to cope with the pulse lengths. This is due to the large values that must be used for both the timing capacitor and the timing resistor, and if we consider the maximum delay time required in this application, values of 1 megohm and 330 microfarads for R2 and C2 would give approximately the required 360 second output pulse. While these values are perfectly satisfactory in theory, in practice a 330 microfarad capacitor would have to be an electrolytic type and would almost certainly have a substantial leakage current. With a 1 megohm timing resistor the charge current would only be in the region of a few microamps, and the leakage current could well be comparable to this. At the very least this would give poor accuracy and unreliable results, and in an extreme case would prevent the circuit from working at all.

The only practical solution to this problem is to use a tantalum bead capacitor for C2. Tantalum capacitors generally have much lower leakage levels and closer tolerances than electrolytic types, but the highest values generally available is only 100 microfarads, and this gives a timing resistance of about 3.3 megohms for an output pulse of about 360 seconds in duration. Using a higher timing resistance reduces the charge current to an even lower value, but 3.3 megohms plus a 100 microfarad timing capacitor are just about tolerable in practice.

One way of obtaining twelve switched times would be to have twelve switched preset resistors in the R2 position, with the resistors being trimmed in value to give precisely the required output times (or reasonably accurate times anyway). This system is perfectly feasable, but does have one slight disadvantage in that the business of adjusting the twelve preset resistors could be rather time consuming and a little tedious. It would be better to use a series of twelve fixed value close tolerance resistors with one preset being adjusted to trim all twelve times to reasonable accuracy. The timing capacitor could be replaced with a set of switched capacitors of course, but this is not a very practical proposition since it would be very expensive, and capacitors of the required value having a close tolerance are not available.

With the 555 it is possible to trim the output pulse length without varying either the timing resistance or capacitance, and all that is required is a potentiometer with its track connected across the supply lines and wiper terminal connected to pin 5 of the 555. This function is performed by VR1 in Figure 2, and what this component actually does is to modify the voltage on C2 at which the timing period is brought to an end. As mentioned earlier, this voltage is normally about two thirds of the supply voltage, but VR1 can be used to reduce this potential so that a shorter output pulse is obtained, or to increase it so that C2 takes longer to reach the threshold level and a longer output pulse is produced.

Alternatives

Although the circuit of Figure 2 could be made to suit our requirements reasonably well it is usually worth exploring a few other possibilities before finally going ahead and trying a circuit in practice. Apart from the possibility that these explorations might turn up a much better solution to the problem, it also gives some alternative approaches that can be tried if the original idea fails to give satisfactory results in practice.

The 555 is not the only timer IC, and there are alternative devices such as the ZN1034E and the UA2240. These are both somewhat more complex devices which use a C – R oscillator plus a divider chain and logic circuitry to give very accurate timing even over long periods. If the 555 circuit was to give inadequate accuracy or prove to be unreliable in practice it would be an obvious next step to try one of these two ICs as an alternative. However, on the face of it the added expense and complexity of these two devices is not justified in this application and these only need to be considered as stand-bys for the time being.

Another possibility would be to use an operational amplifier in a simple timing circuit, and a possible basic configuration is shown in Figure 3. Here the operational amplifier is being used as a voltage comparator with a reference voltage fed to one input of the device and the C – R timing network feeding into the other input. At switch-on C1 is uncharged and the inverting (–) input of the operational amplifier is at zero volts. The non-inverting (+) input is biased by the potential divider formed by R2 to R6, and receives a bias voltage that depends on the setting of S1. For the sake of clarity only four switch positions and five resistors are shown in the diagram, but in practice

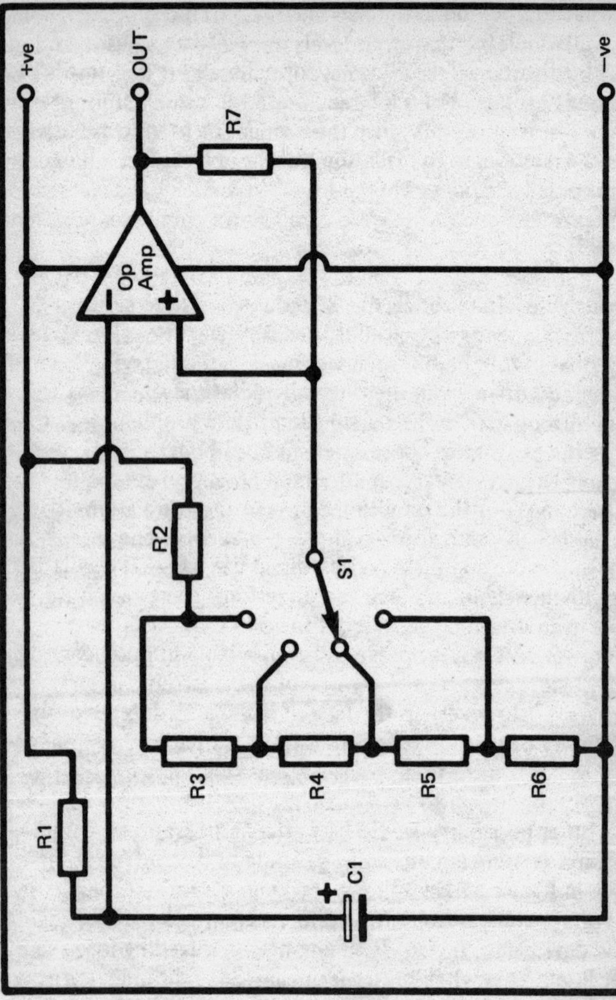

Fig. 3. An Op amp timer circuit

there would be twelve switch positions and thirteen resistors in the chain to give the unit its twelve delay times.

The operation of the circuit is very simple, and the output pulse from the operational amplifier ends when the charge potential on C1 exceeds the bias voltage fed to the non-inverting input. The comparative input levels reverse at this point so that the output of IC1 starts to go negative, and R7 is a high value resistor which is included to give the circuit a small amount of positive feedback so that once the output starts to go negative it rapidly switches to the fully negative state. A slow transition would be undesirable as it would tend to reduce the accuracy of the unit, and could cause the audio alarm circuit to malfunction in some way.

At first sight it might seem that the potential divider chain could be made up from a series of equal value resistors so that the voltages provided by the chain would give a linear series of values, and R1 could then be a preset component to trim the output times of the circuit to give reasonably accurate results. This arrangement would not work in practice since C1 charges in an exponential manner and not in a linear one. In other words the potential on C1 rises relatively fast initially, and then rises at a slower and slower rate. In order to obtain linear increments in the output pulse length the resistor chain would have to be made up from resistors of different values, and even if the required values could be calculated reasonably easily a set of non-standard values would be required.

A simple solution to this problem would be to replace R1 with a constant current source so that the potential on C1 would rise at a steady rate. Another problem with the circuit is that C1 is not discharged at the end of the timing period, and it does in fact continue to charge. This is only of importance in that it could result in a charge being left on C1 if the timer is switched off at the end of a timing run and then switched on again soon afterwards. It would not be too difficult to add a discharge circuit for C1, and this could simply be achieved using an additional pole on the on/off switch to discharge C1 when the switch is in the off position (with a series resistor being used to limit the discharge current and prevent the switch from being damaged).

The configuration of Figure 3 looks as though it is perfectly feasable provided a few minor alterations and additions are made, but it is not necessarily a better one than the simple 555 circuit of Figure 2. It is certainly more novel and interesting, and you may well decide that circuits of this type are what interest you and that circuits do not need to be designed and selected on purely practical grounds. This would not normally be an acceptable working practice for a professional circuit designer, but if you are designing and building circuits for your own amusement it is obviously a perfectly acceptable approach.

However, in most cases it is the more practical circuits that have to be used simply in order to keep the cost and complexity of projects within reasonable bounds. With very simple projects there is plenty of scope for experimentation with different ideas, but with more complex types of project a more rational approach is advisable, otherwise the project can grow to mammoth proportions and become totally impractical.

The configuration of Figure 3 is obviously somewhat more complex than that of Figure 2. Bearing in mind that the operational amplifier in Figure 3 would need to be a high input impedance (FET) type and therefore one of the more costly devices, it would also be somewhat more expensive. In terms of accuracy the configuration of Figure 3 does not seem to offer any advantages, and if anything it would probably be slightly worse since the constant current generator would not have perfect linearity and simple circuits of this type tend to drift slightly but significantly with changes in temperature. The 555 circuit therefore appears to be the more practical solution to the problem, and it seems unlikely that it can be bettered. We will accordingly use the circuit of Figure 2 as the basis of the timer section in this demonstration design.

Alarm Oscillator

There are many oscillator configurations which can be used to provide an audio tone, but many are not ideally suited to this application. The main requirement of the oscillator is that it should be simple to interface to a loudspeaker of some kind so that a sound of adequate volume is generated, and it is essential that the circuit used is one which can be frequency modulated. The output waveform is not really significant in this application, and a simple squarewave or pulse waveform is quite acceptable. It would be useful if the oscillator could be gated by the output of the timer circuit, and would have a negligible stand-by current when gated off so that the battery would not be drained unnecessarily, but this is not absolutely essential as the 555 can control quite respectable output currents without any buffering. It could therefore be used to simply supply power to the oscillator while its output is in the low state (at the end of the timing run) and cut off power to the oscillator while its output is high.

An obvious choice for use as the oscillator is another 555, but used in the astable mode. Figure 4 shows the basic 555 astable configuration, and this operates by first charging C1 to two thirds of the supply voltage through R1 and R2, then discharging it to one third of the supply voltage through R2 and an internal transistor of IC1, then charging it through R1 and R2 again, and so on. The output at pin 3 is high while C1 is charging and low while it is discharging.

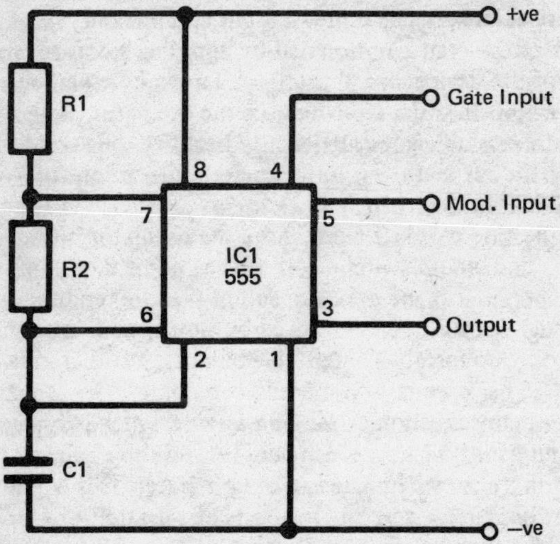

Fig. 4. The basic 555 astable configuration

The output waveform at pin 3 is virtually a squarewave if the value of R2 is high in relation to that of R1 (so that the charge and discharge resistances are almost the same), or a brief negative pulse if the two resistors are of similar values or R1 has the higher value. In either case the output signal is suitable for our purposes, and a further point in favour of the 555 is that it can directly drive a loudspeaker and will give adequate volume for this application.

It is also possible to frequency modulate the 555 by applying a control voltage to pin 5 of the device. This is similar to the technique of trimming the output pulse length of a 555 monostable, but in this case a higher control voltage increases the charge time of C1 and gives a reduction in frequency while a lower control voltage decreases the charge time and gives an increase in frequency. In fact both the charge and discharge times are stretched or shortened, and the frequency modulation does not significantly affect the output waveform, although this is not of any practical importance in the present application.

A 555 astable can also be gated by applying a control signal to pin 4 of the device, and this terminal must be taken below about 0.5 volts in order to switch off the oscillator. The output of a 555 monostable goes sufficiently negative to gate a 555 astable to the off state, but unfortunately the output of the monostable is normally high and goes low at the end of the timing period. In other words it would have the oscillator normally oscillating and would switch it off at the end of the timing period, whereas it is the opposite action that we require!

This slight problem can be cured by simply adding an inverter between the monostable and astable circuits, but there is a more serious drawback which is not easily overcome. This is merely the current consumption of the 555 astable which is typically about 8 milliamps using a 9 volt supply, and the oscillator would consume this even when gated to the off state. This would significantly increase the running costs of the circuit by necessitating the use of a large 9 volt battery to power the unit, or giving a very restricted battery life if a small type such as a PP3 was to be used. There is a low power CMOS version of the 555 called the ICM7555, or sometimes it is referred to just as a 7555. However, this device has an output stage which has neither the source current nor the sink current capability of the standard 555, and it gives substantially less volume when used to drive a loudspeaker.

In summary, the 555 is not perfect for this application, but it is certainly usable and would suffice if nothing better could be found.

CMOS Oscillators

Other obvious possibilities for the audio oscillator are circuits based on operational amplifiers, discrete transistors, or CMOS logic ICs. The latter seems to be the most promising in that gating can be easily achieved, and CMOS circuits usually have a very low level of stand by current consumption.

There are two CMOS ICs which are specifically designed for oscillator applications (amongst other things), and these are the 4046 and 4047 ICs. The 4047 can be used as a gated oscillator but has no input to permit frequency modulation. The 4046 is actually a phase locked loop integrated circuit, but it can be used as a voltage controlled oscillator if the phase comparators in the device are simply left unused. Being a voltage controlled oscillator it could obviously be frequency modulated without difficulty, and it can also be gated. This device would therefore seem to be a reasonable choice.

CMOS oscillators can also be built using simple inverters or gates connected to provide an inverter action. Figure 5 shows the basic configuration used for this type of oscillator, and it has been assumed here that two input NAND or NOR gates are used as the active elements of the circuit. The gates used can have any number of inputs, and all the inputs of each gate are connected together to give an inverter action, but AND, OR and XOR gates cannot be used as they would not give an inversion from the input to the output. R1 biases the two inverters (which are wired in series) so that they act as a linear amplifier, and C1 provides positive feedback which causes the circuit to oscillate strongly. A roughly squarewave output is produced by this circuit.

Using this simple oscillator configuration is an attractive proposition because a single CMOS IC could provide both the

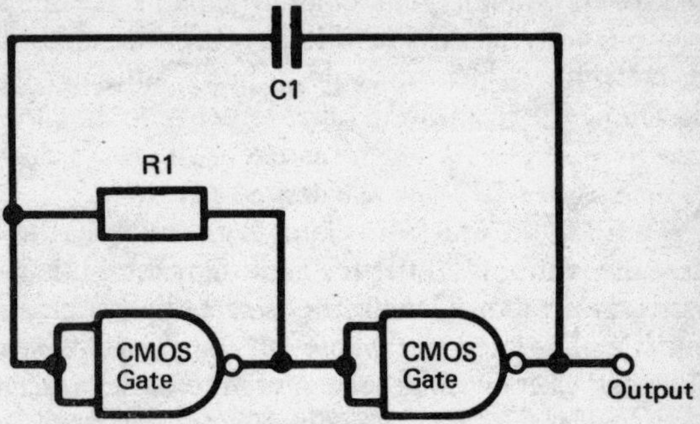

Fig. 5. A simple CMOS Oscillator circuit

audio and low frequency oscillators at a cost that would be lower than any alternative (as far as I am aware anyway). The obvious problem is that the basic CMOS astable configuration does not provide modulation or gating inputs.

Using gates as the basis of the oscillator it is quite easy to obtain a gating action, and it is merely necessary to take the gating signal to one input of either gate. When this input is taken to one logic state the inverter action will be provided and the circuit will function, but with it taken to the other logic state the output of that gate will be held at a fixed logic level and the action of the circuit will be blocked. The logic level needed to block operation and the output state obtained while the circuit is blocked depends on the type of gate used.

It would be wrong to assume that there was no way of modulating the oscillator simply because the basic circuit does not provide a suitable input. A little experimentation with a control voltage being loosely coupled to various parts of the circuit could show that the operating frequency can be modulated. The circuit is so simple that it could be quickly built up using one of the solderless breadboards that are readily available these days, and a suitable control voltage could be produced by connecting the track of a potentiometer of a few kilohms in value across the supply rails with the variable voltage being extracted from the wiper terminal.

On trying this it was found that loosely coupling the control voltage to the input via a high value resistor (about double or more the value of R1) gave the desired effect.

The output of a CMOS oscillator is not sufficient to drive a loudspeaker, but this is not very important since it is quite easy to boost the output of the circuit using a single transistor amplifier, and the additional cost and complexity is negligible.

We can therefore achieve the desired result using a circuit of the type outlined in Figure 6. The circuit could be based on NAND or NOR gates, but not inverters which would not permit the circuit to be gated. NOR gates were chosen simply because there were several suitable devices in the author's spares box, and there is no point in buying components if you already have suitable devices. The ability to use up "left overs" is one of the advantages of designing your own projects.

The circuit is very much along the lines outlined above with the modulation applied to the input of the first gate via R3. The output of the circuit drives a loudspeaker via a single transistor (Tr1) common emitter amplifier with the loudspeaker being directly driven by this. R2 is needed primarily to prevent the base-emitter junction of Tr1 from acting like a low voltage zener diode and preventing the output of the oscillator from reaching a potential of more than about 0.65 volts. This would almost certainly prevent the oscillator from operating at all.

The gate signal is applied to one input of the second gate. The output of a 2 input NOR gate is low if either of the inputs is taken to the high logic state. Thus if the gate input is taken high the output of the circuit will go low, and the logic level at the other input of the gate is irrelevant. This blocks the action of the circuit and no output is obtained. If the gate input is taken low, the logic level at the output of the gate depends on the input level at the other input, and the circuit can therefore function normally. Thus the required gating action is obtained

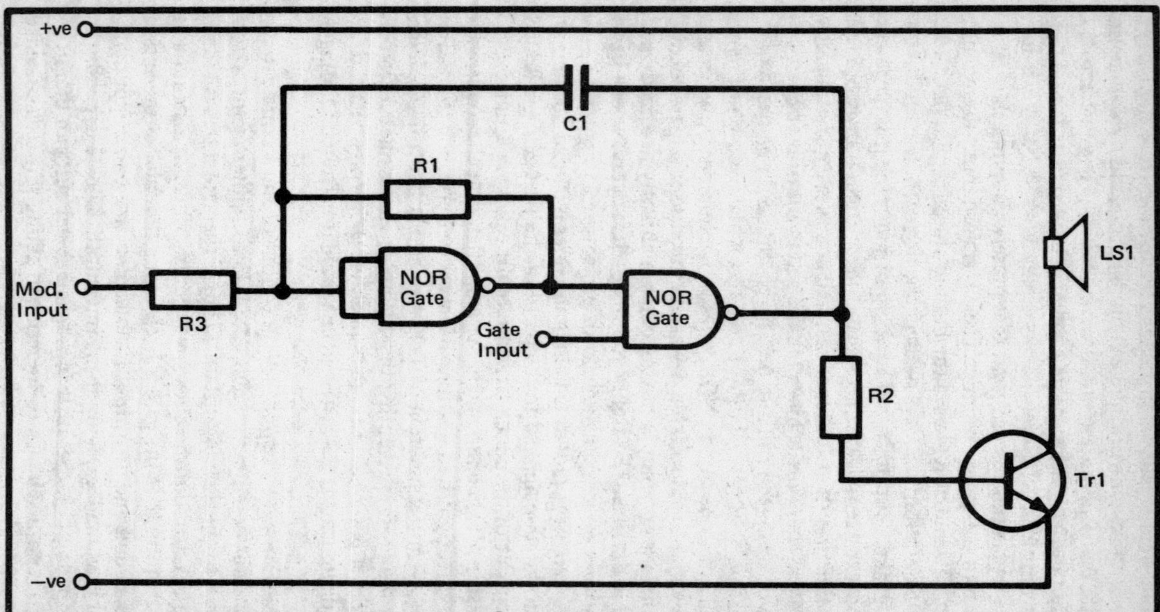

Fig. 6. A CMOS tone generator with Mod. and Gate Inputs

with a low control signal producing oscillation and a high one blocking the circuit, and there is no need to add an inverter at the gate input.

An important point to note is that the output of the circuit is low when the gate input blocks oscillation. Tr1 is therefore cut off and no significant current is passed by this device. If the output went high when oscillation was blocked Tr1 would be biased hard into conduction and the circuit would have a high current consumption. When designing circuits it is essential to look carefully for points of this type so that they can be corrected if necessary at an early stage. With this configuration the output of the oscillator can be low or high, depending on whether it is the first or second gate to which the gating signal is applied.

Final Circuit

Putting together the various ideas discussed so far, the final circuit of Figure 7 was produced. R3 is shown as a single resistor in Figure 7 for the sake of clarity, but in practice this would be a series of 12 switched 330k resistors, as shown in Figure 8.

The overal operation of the circuit should be clear to you, since IC1 is a straight forward 555 monostable, IC2a and IC2b form the tone generator which drives the loudspeaker via amplifier Tr1, and the other two gates of IC2 form another oscillator which is used to provide the low frequency modulation by way of R6. The way in which the various circuit values were determined will probably not be apparent, and we will therefore consider each component in turn. However, an important point to make before we start this, is that there is usually considerable latitude when it comes to choosing component values. There are times when a circuit value is very critical, but in many cases changing a component value by as much as plus 100% or minus 90% will have no significant effect on performance. In many cases the value of one component is only critical in relation to that of another. In other words changing the value of one component will not upset the operation of the circuit provided the value of another component is changed accordingly. A simple example of this is the timing components for a monostable or astable circuit. If the value of the timing resistance is

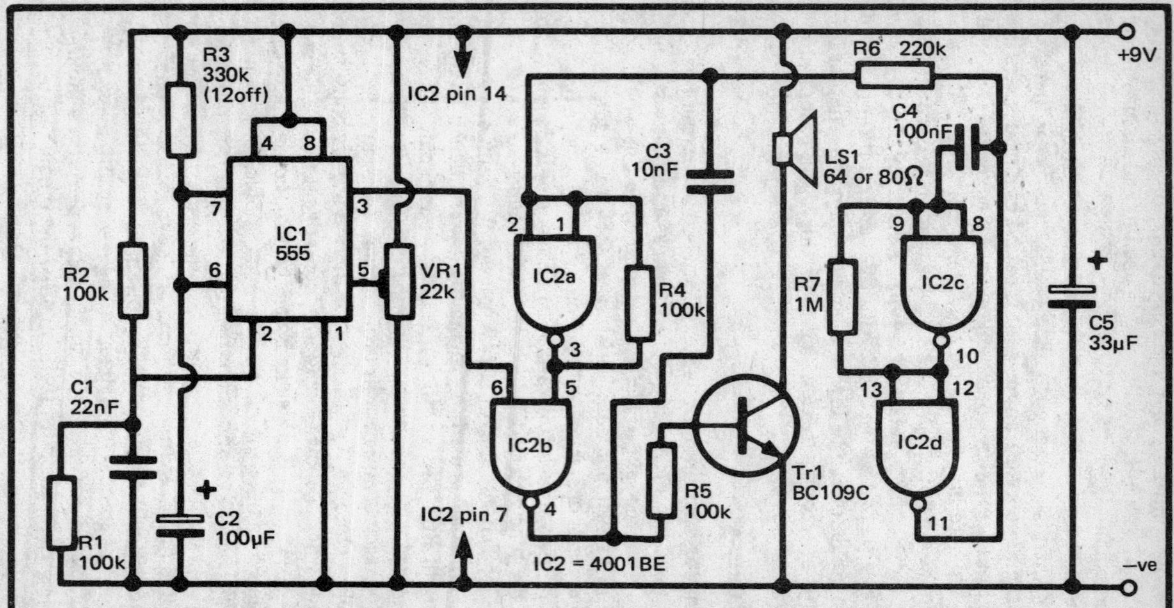

Fig. 7. The final Timer circuit

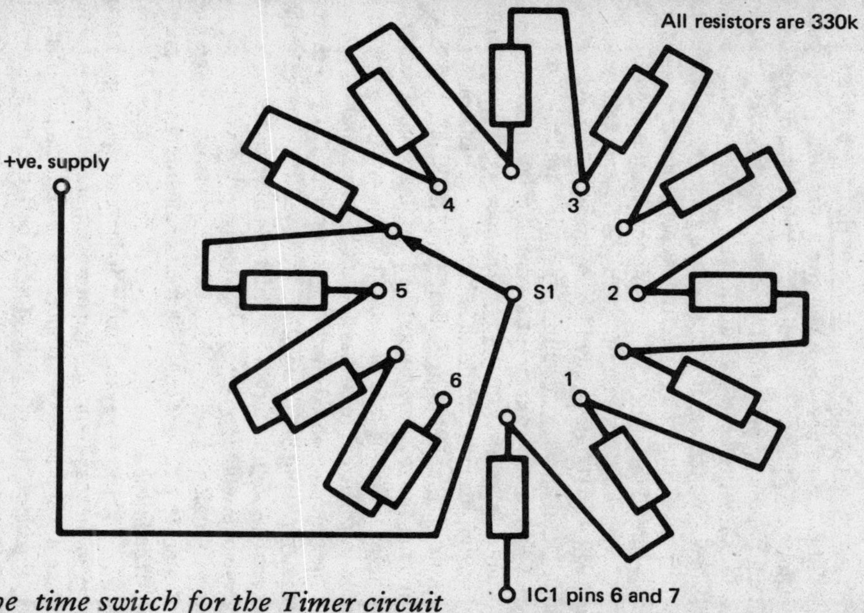

Fig. 8. The time switch for the Timer circuit

increased there is an increase in the output pulse duration or a decrease in the output frequency (depending on the type of multivibrator in use). Reducing the value of the timing capacitance in proportion to the increase in the timing resistance returns the original time constant, and hence the original output pulse length or frequency. This means that there are a wide range of C – R values which will give a specified output pulse duration or frequency, but only one timing capacitance that will suit a given timing period or frequency for a specific resistance. It is important to be able to distinguish between critical and non critical components as this can avoid a great deal of wasted time, and the possibility of a perfectly feasible project being abandoned simply due to the value of one component not being quite right.

The value of C2 is a point which was covered earlier, and this has a value of 100μF simply because this is the largest value in which tantalum bead capacitors are readily available. It would be possible to use a somewhat lower value, but it cannot be substantially lower as the timing resistance required for the six minute output pulse would then be too large. It is advisable to keep the maximum timing resistance for the standard 555 to no more than about 10 megohms (20 megohms is the absolute maximum). On the face of it an operating voltage of 6 volts would be sufficient for C2 as a 9 volt battery supply is used, and the potential on C2 reaches a nominal level of two thirds of the supply potential (6 volts). However, this overlooks the fact that a 9 volt battery when fresh can have an actual output voltage of around 9.5 volts. Also, the two thirds of the supply voltage threshold level at which C2 becomes discharged is only a nominal level, and it could be fractionally more in practice. In this circuit this threshold voltage might be boosted by VR1 anyway, and it is always advisable to leave at least a small margin between the maximum voltage of a component and the highest voltage it will ever receive (especially in a critical part of a circuit as in this case). It would therefore be better to use a component having a somewhat higher maximum working voltage in the C2 position, and a 10 volt component would be suitable. In fact any voltage of more than about ten would be suitable, but in practice a ten volt component is all that is likely to be available.

The value of the resistors in the R3 position must be chosen to give an output pulse length of about 30 seconds per resistor, and as mentioned earlier, the output pulse from a 555 monostable is nominally 1.1CR seconds. With a 100μF timing capacitor this gives a timing resistance of a little under 300k, but due to the inclusion of VR1 the actual value used does not have to be particularly close to the calculated figure, and there is a latitude of about plus or minus 50% here. The value of 330k was selected as suitable components were to hand, but obviously 270k, or 300k components would be equally suitable. Note though, that all twelve resistors should have the same value, and that these should ideally be close tolerance (1% or 2%) types. This is to ensure that with VR1 adjusted to give a delay of thirty seconds with S1 at the first position, good accuracy is obtained at all the other switch positions with the delay time going up in 30 second increments.

The value of VR1 is not too critical, and it is really just a matter of not choosing a value that is so low as to substantially add to the current consumption of the circuit, or so high that it is unable to pull the threshold voltage of IC1 up and down in the required manner. Of course, however high the value of VR1 is made it can still connect pin 5 of IC1 to either supply rail, and therefore provides full adjustment range. The problem with using a high value is really that it would tend to have practically no effect over the middle 90% or so of its track, but would instead provide almost the full adjustment range at the ends of the track. This would make accurate adjustment difficult, and there is also the danger that an excessive current would be passed through the track when it was adjusted towards one end or the other.

A value of 22k adds under 500 microamps to the current consumption of the circuit but gives more than adequate current to drive pin 5 of IC1 properly. In practice any value of between about 4k7 and 100k would probably be perfectly satisfactory.

The values of R2 and C1 are not at all critical. The 555 needs a trigger pulse of at least 0.1 microseconds, and the trigger pulse must not be so long as to retrigger the circuit when the timing period ends, but this obviously gives a very wide latitude indeed. It would not be a good idea to use a very short trigger pulse as it would then be possible for the voltage at pin 2 of IC1 to lag only marginally behind the supply voltage as this builds up (it does not rise instantly to 9 volts due to the inclusion of supply decoupling capacitor C5), and the circuit would not be triggered at all. On the other hand there is no point in having a very long trigger pulse with a large and expensive capacitor being used in the C1 position. Another point to bear in mind is that if R2 is made very low in value the trigger circuit will load the supply very heavily at switch on as a large charge current flows into C1, and this could cause a malfunction of the circuit.

The specified values avoid all the pitfalls mentioned above, but if R2 and (or) C1 were to be raised or reduced in value by a factor of ten the circuit would almost certainly still function perfectly.

R1 was not originally included in the circuit, and was added when a minor fault in the unit became evident. It was found that if the unit was switched on again soon after the end of a timing run the alarm was activated immediately. This tended to suggest that IC1 was not being triggered, which in turn suggested that C1 was not discharging rapidly enough at switch-off. Adding R1 limited the maximum charge on C1 to half the supply voltage and also provides a discharge path for C1 when the unit is switched off. The addition of R1 seemed to complet eliminate the problem but did not have any adverse effect on the circuit.

It is difficult to adopt a highly scientific and technical approach to choosing the values of the timing components in the audio and low frequency oscillators. One problem is that component tolerances, particularly the tolerances of the NOR gates, makes the exact operating frequency of the circuit for a particular set of values slightly unpredictable. The main problem though, is simply that the circuit will provide perfectly satisfact results with the tone oscillator at any frequency from a few tens of Hertz to a few kilohertz, and the low frequency oscillator could have any frequency from about 0.5 Hertz to around 10 Hertz.

Similarly, the depth of the modulation is controlled by the value of R6, and you might like to use a high value here (with little modulation) or a low value (and large modulation). As mentioned earlier, the only restriction on the value of R6 is that it should not be much less than about double the value of the timing resistor used in the tone generator.

The values used in the alarm generator must be chosen to suit the particular alarm sound that is required, and this is really a matter of experimenting with various combinations of tone frequency, modulation frequency, and modulation depth, to see what combination you consider to be the most suitable. Alternatively, the circuit values could simply be calculated to give frequencies and modulation depth that would be almost certain to give satisfactory results in practice, and you would then accept whatever alarm sound the unit happened to produce (although it could be modified if necessary). For example, a nominal frequency from the tone generator of about 800Hz (i.e. a middle audio frequency) with a modulation frequency somewhere near the middle of the acceptable frequency range (around 2 or 3Hz) and a moderate amount of modulation (R6 at about four or five times the value of R4) would almost certainly give a quite effective and acceptable output.

Whether the circuit values are calculated or found by experiment, there are a few limitations which must be kept in mind. The value of the timing resistor in each oscillator should not be less than a few kilohms in value or the input impedance to the first inverter of the oscillator becomes so low that the circuit may not function properly. The upper limit for this resistor is set at about 10 megohms, and this is merely because higher value resistors are not readily available. It is advisable not to use a timing capacitance of less than about one hundred picofarads as this could result in the circuit failing to oscillate, although it would not normally be necessary to use such a low value anyway. In theory there is no upper limit to the timing capacitor's value, but as it cannot be a polarised component such as a tantalum bead or electrolytic type, the maximum practical value is about 2.2 microfarads. It is usually better to use a high value resistor rather than a high value capacitor when a low output frequency is needed as this avoids the expense of a high value non-electrolytic capacitor.

The operating frequency of this type of oscillator is roughly 1/(1.4CR) Hz incidentally, but the circuit values used in the two oscillators of Figure 7 were found by experimentation.

Probably the easiest component selection is choosing the device for the IC2 position. The two oscillators require a total of four NOR gates, and looking down the list of CMOS logic devices the 4001BE quad 2 input NOR gate IC is almost the first on the list (to be precise, its the second). This is obviously ideal for our purposes and there is not really any point in searching further.

The choice of the device used in the Tr1 position is equally easy, but for a different reason. Practically any NPN silicon transistor could be used here, and the only requirements are that the device should have a reasonable current gain (about 50 or more) and that it should be capable of handling a reasonable collector current (say about 100mA or more). There are probably thousands of transistor types which fit these simple requirements.

I keep a stock of BC109C (NPN) and BC179 (PNP) transistors as these cover most requirements, and I also have a stock of more specialised devices for applications where these will not suffice, such as power transistors for high current or high power applications. In fact most circuit designers and experimenters keep a stock of fairly inexpensive and versatile devices such as the BC109C and BC179 and use the relevant type in any circuit where the device could reasonably be used. There is little point in keeping a stock of numerous transistor types which in practice are largely interchangeable and just a couple of types would be almost as good. It is definitely advisable to have a fairly substantial stock of resistors, capacitors, and the other types of component that are in frequent use as this can avoid a lot of waiting for "out of stock" components to be delivered before you can try out your latest idea. It is also a great advantage to be able to try out ideas while they are fresh in your mind rather than a few weeks later when all the necessary components have been assembled. Obviously there will be many occasions when a special component or components are required, but having a stock of the more commonly used components will help to avoid any unnecessary delays in trying out a circuit idea.

As most readers will have gathered by now, the BC109C is used in the Tr1 position simply because this is the device which I normally use when a general purpose NPN transistor is required.

The BC109C (and similar devices such as the BC239C and BC650) have high gain, low noise, and useful maximum collector current and collector-emitter voltage ratings. They are therefore usable in some critical applications such as low noise audio amplifiers, and can even be used in some RF applications. However, if you feel that some other device would be more versatile, then by all means use this instead.

The output of a CMOS logic IC, provided it is not heavily loaded, swings between virtually the full positive and negative supply voltages. Tr1 is therefore fully cut off when the output IC2b is low, and it could be driven to saturation when this output is high, but only if R5 is given a low enough value. In this case it is essential that Tr1 should not be fully switched on since this would cause a current of more than 100 milliamps to flow in its collector circuit. Apart from the fact that this is higher than the maximum permissible collector current for the BC109C device, it would probably give an output power to LS1 that would be well in excess of its maximum rating, and it would give the circuit a rather high current consumption when the alarm was operating.

One solution to this problem would be to add a resistor in series with LS1 to give a total collector load impedance of about 95 to 100 ohms. This would ensure that even with Tr1 biassed hard into conduction an excessive collector current could not flow. A simple alternative, and the method used here, is to limit the drive current to Tr1 to a level that cannot produce an excessive collector current. Looking at the data for the BC109C it is clear that with a large collector current flowing and only a small collector voltage (remembering that as the collector current increases the collector voltage will fall) the current gain of the device will be less than 900. Even looking at the brief data for the BC109C one could deduce this because the brief data shows that the device has a maximum current gain of 900 with a collector voltage of 5 volts and a collector current of 2 milliamps. The gain of a transistor tends to rise with increasing collector current, but only up to a certain point, and above this point the current gain reduces. Current gain also tends to reduce slightly with falling collector voltage. It is therefore unlikely that the BC109C would produce a current gain of more than 900 when fully driven in this circuit, but we can always leave a safety margin if in doubt, and we do not really need to have the collector current close to the 100 milliamp maximum anyway. A lower figure of around 50 milliamps should give a perfectly adequate volume level for this application.

One way of determining a suitable value for R5 would be to assume that the maximum gain of Tr1 will be no more than 1000 (which leaves a safety margin), and to assume that a maximum collector current of 80mA is required. This last figure is low enough to ensure that Tr1 and LS1 are safe, the circuit has a reasonably low current consumption, but if Tr1 happens to be a fairly low gain example of a BC109C (with a gain of around 400) it will still produce a reasonably high current through the loudspeaker and give adequate volume. Dividing 80 milliamps by 1000 gives the required base current into Tr1, and this is obviously 0.08 milliamps, or 80 microamps in other words. The voltage across R5 is equal to the 9 volt supply potential minus about 0.65 volts dropped across the base-emitter junction of Tr1, and is therefore nominally 8.35 volts. Using Ohm's Law we can calculate the value of R5 as 8.35/0.08 = 104.375 kilohms. Note that the answer to this calculation is in kilohms and not ohms since we are dividing 8.35 volts by milliamps rather than by amps. This is generally a quicker and easier way of tackling this type of calculation. If the current figure used in this type of calculation is in microamps, then the answer is in megohms.

104.375 kilohms is clearly not a preferred value, but it is simply a matter of choosing the nearest preferred value which will be 100k in this case. Trying out the circuit using a 100k component and plugging a variety of BC109C transistors into the circuit in turn always seemed to give satisfactory results with an acceptable volume level.

The choice of loudspeaker impedance and size is generally quite easy, although it is often necessary to make a compromise due to the lack of an ideal component for a particular application. This is an example of a circuit where such a compromise has had to be made. For optimum efficiency we require a loudspeaker that will give the desired output power if it is fed with a peak to peak voltage roughly equal to the supply voltage. The RMS output power fed to a loudspeaker can be calculated by squaring the RMS output potential (in volts) and dividing this by the loudspeaker impedance (in ohms) to give an answer in watts. This system is really intended for circuits where an AC signal is applied to the loudspeaker, and a sinewave test signal is used. It is difficult to realistically relate this circuit to the formula given above since the signal is a squarewave and it is applied to the loudspeaker as a pulsed DC signal.

However, we can simply take the power fed to the loudspeaker when Tr1 is switched on, and then divide this by two to compensate for the fact that Tr1 is switched off for about 50% of the time. This gives a reasonably accurate figure for the actual power fed to the loudspeaker, but due to the nature of the signal it gives a slightly optimistic impression of the volume that will be obtained. Even so, in order to give an output of adequate volume to be readily noticed in an ordinary domestic environment we do not require a very large output power. In fact only about 50 to 150 milliwatts is adequate, and many miniature loudspeakers have a maximum power rating of only 100 or 200 milliwatts anyway.

To find the correct speaker impedance we first need to determine the current required to give the desired output power, and this is obtained by dividing the required power (say 0.1 watts or a 100 milliwatts in other words) by the supply voltage. 100 divided by 9 is 11.111 milliamps (the answer is in milliamps rather than amps because the output power was entered in milliwatts and not watts). To compensate for the fact that Tr1 is switched off for about 50% of the time we must double this figure so that an average of 11.111 milliamps is fed to the loudspeaker. We therefore require a current of 22.222 milliamps.

We can now use Ohms Law to calculate the required loudspeaker impedance, and this is 9 divided by 22.222 equals 0.405 kilohms, or 405 ohms in other words. This is not possible in practice as loudspeakers having such a high impedance are not available. Unfortunately there are no loudspeakers which even come close to our requirements, and the best that can be done is to use a 64 to 80 ohm component (the highest available impedances). However, due to the lower speaker impedance it is necessary to add some form of current limiting to the circuit to prevent an excessive output power from being obtained. As pointed out earlier, this can be achieved by adding a resistor in series with the loudspeaker, or preventing Tr1 from switching fully on so that this effectively adds a suitable resistance in series with LS1.

This is not really a good way of doing things since there is a substantial amount of power wasted in either Tr1 or the series resistor, depending on which method is used. This is the reason for the circuit having a loudspeaker current which is typically more than double the 22.222 milliamps we calculated earlier. The extra current is needed to compensate for the drop in drive voltage to LS1 and the consequent loss of output power, and the price that is payed for the lack of efficiency is a higher battery drain than is really necessary.

An ordinary 8 ohm impedance loudspeaker would obviously be of little use in this application. The voltage developed across the loudspeaker would be small while that developed across Tr1 would be large, giving most of the output power in Tr1 and little in LS1.

The size of the loudspeaker used depends largely on how large or small the finished unit is going to be. In this case it would presumably be constructed as quite a small unit and a loudspeaker of no more than about 76mm in diameter would have to be used. In general the quality and efficiency of loudspeakers increases with the size of the component, but obviously in this application high efficiency and good quality are of no significance anyway.

Supply Decoupling

The purpose of supply decoupling capacitors is something that mystifies many beginners at electronics, and on the face of it there is no point in having a capacitor wired across the supply lines, but this is something that can be found in practically any piece of electronic equipment. The purpose of these capacitors is simply to smooth out any variations in the supply voltage that occur due to variations in loading.

There may be no obvious reason for the supply voltage varying with changes in load current, but if you take a small 9 volt battery such as a PP3 (or equivalent) and monitor the output voltage while connecting various load resistances across the output terminals, you will find that the higher the load current the lower the output voltage obtained. This loading of the output voltage tends to be much more noticeable with a nearly exhausted battery than with a fresh one.

This reduction in supply voltage is due to the internal resistance of the battery, but as far as its practical effects are concerned it is much the same as if there was an external series resistor in the output lead from the battery. The internal resistance of a battery varies considerably from one type to another, and also varies according to the amount of charge remaining in the battery. For a small 9 volt battery such as a PP3 it is likely to be between ten and twenty ohms when the battery is fresh, and probably around forty to fifty ohms when the battery is nearing the end of its useful life. For larger 9 volt batteries such as PP7 and PP9 types the internal resistance is significantly less, and for something like a single high capacity NiCad cell it would be only a fraction of an ohm.

If a circuit has a constant current consumption there is little point in using a supply decoupling capacitor. The battery voltage would drop when it was initially connected to the circuit, but the supply voltage would then remain constant. This is quite rare though, and with most circuits there are significant changes in the current consumption during the normal operation of the unit, and in many cases there are very large changes in the current consumption.

This circuit is no exception, and the current consumption is obviously very much higher during the periods when Tr1 is conducting than when it is switched off. This rapid modulation of the supply voltage could result in a malfunction of some part of the circuit, although this is not likely in this case. The oscillators do not start to function until IC1 has finished providing an output pulse, and the only way in which IC1 could be caused to malfunction would be if it was to be retriggered. There is no obvious way in which this could occur. It is unlikely that a serious malfunction in either of the oscillator circuits would occur either since the configuration used is quite tolerant of supply voltage fluctuations.

The main reason for including C5 in this circuit is that it gives a slightly higher output to the loudspeaker, especially when the battery is nearly exhausted. Remember that the circuit is intended for use with a small 9 volt battery such as a PP3, and that the internal resistance of this will become quite high towards the end of its operating life. Apart from the drop in unloaded voltage that occurs, a further and quite significant fall will occur due to the increased current flow when Tr1 is pulsing a current through LS1, and this gives a low supply voltage at a time when a high supply voltage is needed to ensure that LS1 is driven properly. Reducing R5 to a much lower value would give a good drive level to LS1 with an old battery in use, but it would also give an excessive drive current with a fresh battery in use.

Adding a supply decoupling capacitor is a better solution, and this capacitor charges during the periods when Tr1 is switched off, and gives up some of this charge while Tr1 is conducting so that the fall in supply voltage is minimised. This helps to give results that are as consistent as possible during the operating life of each battery. Also, as the battery voltage drops and its internal resistance rises it is likely that a point will eventually be reached where the circuit does malfunction in some way, rather than the volume of the alarm signal gradually falling until it becomes inadequate. Using a supply decoupling capacitor helps to put off this eventual malfunction of the circuit for as long as possible so that as much power as possible can be extracted from each battery.

The value needed for a supply decoupling capacitor is something which is dependent on a number of factors. One of these is the source impedance of the power supply, and the higher this impedance the greater the supply decoupling capacitor that is likely to be needed. Another is the variation in supply current, and to an extent the average level of current consumption is important. The greater the current consumption and the supply current variation the higher the value of the required supply decoupling capacitor. The minimum frequency handled by the circuit (and therefore presumably the minimum frequency of supply current fluctuations) is also crucial. If the supply current variations occur at a high frequency the supply decoupling capacitor only has to provide very brief pulses of current during peaks of current consumption, and a relatively low value can therefore be employed. If the variations have a frequency of as little as a few Hertz, peaks in current consumption will last as much as a few hundred milliseconds and a large decoupling capacitor is then required.

As a general rule of thumb, a decoupling capacitor of about 10nF in value is suitable for VHF circuits, 100nF is suitable for RF circuits up to the start of the VHF range (i.e. 30MHz approximately), and about 100μF is suitable for audio frequency circuits. The value of supply decoupling components is not usually too critical, and it does not matter if the value used is somewhat larger than the minimum value that would give adequate decoupling. In fact it is probably advisable to make the capacitor somewhat larger than is really necessary, and it is not usually a good idea to leave no margin for error. However, when dealing with RF circuits remember that high value capacitors, especially electrolytic types, tend to be inefficient at radio frequencies, and by using a decoupling capacitor which is considerably over-size it is possible that inadequate decoupling would be obtained. It is sometimes necessary to have supply decoupling which will be effective at both high and low frequencies, and it is then advisable to connect (say) a 100μF and a 100nF capacitor in parallel across the supply lines. In effect the 100μF capacitor takes care of the low frequency decoupling while the 100nF capacitor handles the high frequency decoupling where the electrolytic will be inefficient and ineffective.

A point which often puzzles beginners at electronics is where an audio amplifier is powered from a well smoothed and regulated supply which has a very low source impedance which for practical purposes can be regarded as zero, and there may also be a substantial smoothing capacitor at the output of the supply. There is also a decoupling capacitor across the supply lines of the audio amplifier, but this is in parallel with the smoothing capacitor at the output of the power supply and is apparently superfluous.

In theory it is superfluous, but in theory there is assumed to be zero impedance in the wiring of the power supply and the amplifier. In practice there is obviously a certain amount of resistance in this wiring, although this resistance is likely to be a matter of milliohms rather than ohms. However, if there is a resistance of (say) 100 milliohms (0.1 ohms) in the wiring and the amplifier is a power type having a peak current consumption of perhaps 3 amps, this gives a voltage drop through the wiring of 300 millivolts (0.3 volts) at times of peak consumption. Most power amplifiers are Class B types these days and these have only a very low current consumption when there is only a small output signal. Thus the supply voltage would vary by nearly 300 millivolts even if the power supply had perfect regulation.

This is obviously a significant variation, but it should be borne in mind that half this variation is in the non-earthy supply lead, but half is in the earth supply lead, which is of greater significance. The audio amplifier is likely to be designed so that it is not much affected by variations in the non-earthy supply (although it is advisable to eliminate these if possible so as to optimise performance and reliability), but variations in the earth supply rail are likely to be fed direct to the input of the amplifier, since one input is almost certain to be connected to the earth rail.

Supply decoupling can help to prevent problems with instability, distortion, or loss of gain due to voltages developed in the supply wiring, but this will not always be a complete answer. The arrangement of the earthing is often a crucial factor, and if this is arranged incorrectly the circuit cannot function properly. There are two basic types of earthing arrangement whic can be used in this situation, "spider" earthing and "bus-bar" earthing. These are illustrated in Figure 9(a) and Figure 9(b) respectively. In both cases the earthing is arranged so that any change in potential at the input of the amplifier is matched by an identical change in the potential at the output so that no potential difference is developed between the two, and unwanted feedback is avoided.

With low level circuits the currents involved are much lower, and any voltages developed in the supply wiring will therefore be much lower as well. However, this is cancelled out to a certain extent by the fact that the input sensitivity of the circuit may

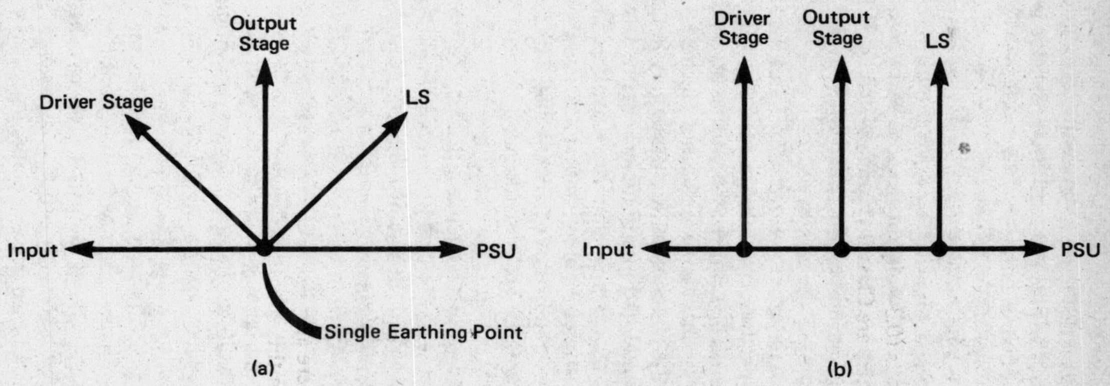

Fig. 9 (a) The "Spider" earth system and (b) the bus-bar earth system

well be very high with an input of under one millivolt perhaps being sufficient to fully drive the circuit. Where a circuit has a combination of high gain and output power it is essential to take great care with the earthing arrangement if instability or poor performance are to be avoided.

In radio frequency circuits, particularly at VHF and higher frequencies, it is the inductance in the wiring that tends to cause problems, and using long leadout wires on a decoupling capacitor can be sufficient to render it useless. However, this is really a rather specialised subject and goes beyond the scope of this publication.

If we return to the circuit of Figure 7 and the value of the decoupling capacitor (C5), this has a somewhat lower value than the 100μF mentioned earlier as a satisfactory value for most audio frequency decoupling applications. This is quite acceptable though since the circuit does not handle very low audio frequencie and the maximum output current and supply current variation are not very large. Higher values were tried out in the circuit and gave no noticeable change in performance even when the circuit was powered from a nearly exhausted PP3 size battery.

When trying out a breadboarded circuit prior to making the circuit properly on a printed circuit board (it is a good idea to always test and correct a circuit in this way before going to the time and expense of producing a printed circuit board where it is feasable to do so), if the circuit is battery powered it is essential to test the circuit using a well used but serviceable battery. The fact that a circuit will function using a brand new 9 volt battery which has a low internal resistance and an actual voltage of about 9.5 volts does not mean that the circuit will work properly using a well worn battery with a much higher internal resistance and an output voltage of about 7.5 volts. Additional supply decoupling may have to be added, or perhaps some component values will need to be changed to suit the reduced supply voltage.

If you use a bench power supply when testing circuits there is a real danger that a circuit which functions perfectly using the bench supply will fail to operate properly when powered from a well aged battery, or even if it is powered from a fresh one. Modern stabilised power supplies have a very high level of performance indeed, and the difference between the loaded and unloaded output voltage is often a matter of just a few tens of millivolts. This represents a far lower internal resistance than is obtained using a normal 9 volt battery, and this can mask problems that will occur when the circuit is used with a battery supply.

The obvious solution is to try the circuit using a battery supply before finalising the design, but there is a simple alternative which is to simply add a resistor in series with the non-earthy output of the supply. This resistor effectively reduces the regulation efficiency of the power supply, and simulates the internal resistance of a battery. For a large 9 volt battery a resistor of about 10 to 15 ohms in value would be a suitable choice, while a value in the region of 39 to 47 ohms would be suitable if the circuit is designed for use with a small battery such as a PP3 type.

Logic Decoupling

Before leaving the subject of supply decoupling it would perhaps be as well to mention logic ICs and a few other devices which have special decoupling requirements. CMOS logic ICs are probably the least demanding, and due to their excellent immunity to noise on the supply rails they require a minimal amount of supply decoupling. It is often possible to obtain perfectly satisfactory results using just a single 100nF supply decoupling capacitor, even in a fairly complex system. In a simple system there would probably be no ill effects if no supply decoupling at all was to be used, but it is probably better to play safe and add a 100nF supply decoupling capacitor to the circuit.

TTL logic ICs (including LS types etc.) have far lower supply noise immunity, and it is normal to have at least one 100nF supply decoupling capacitor for every five TTL devices. These decoupling capacitors should be placed in a well spaced pattern over the circuit board, and not with them all grouped together in one place. With devices such as counters, shift registers, and buffers it is advisable to have a 100nF supply decoupling capacitor for one or two devices, and ideally there would be a decoupling capacitor for each device, and mounted as close as possible to that device.

It is not uncommon for complex integrated circuits to require a supply decoupling capacitor of around 100nF in value mounted as close to the device as possible. Although there is a temptation to think of an IC as a single stage of a circuit, in most cases there are numerous stages inside the device and the decoupling capacitor is needed to prevent instability or a malfunction occuring due to one stage modulating the supply to another stage, just as for a discrete multi-stage circuit. A supply decoupling capacitor of this type is a common requirement for devices such as operational amplifiers, comparators, IF/RF amplifiers, and audio amplifiers. Three terminal monolithic voltage regulators normally require two decoupling capacitors of around 100 to 220nF mounted physically close to the device; one connecting across the input and one across the output.

Batteries

The on/off switch is not shown in the circuit of Figure 7, but this could be any simple on/off type; toggle, miniature toggle, push button, slider, or rotary, as preferred. S1 is simply a 12 way single pole rotary switch, and these are readily available. In fact this type of rotary switch is available in two types from some suppliers. The two types are called "make before break" and "break before make". With the make before break type, if the switch is taken from (say) position 2 to position 3, during the transition there is a short circuit between tags 2 and 3, and between these two tags and the pole terminal of the switch. With the break before make type the opposite occurs with no connection between the pole terminal and any of the other tags for part of the transition between switch positions.

In most applications, including the present one, it does not matter which type is used. However, there are occasions when using a make before break type would not be advisable, and there are circuits where, for example, using this type of switch would cause a brief short circuit across the supply lines during the move from one position to another. Apart from any damage this might cause to the power source, this could seriously reduce the operating life of the switch. With the break before make type such a short circuit could not occur, but the gap in continuity between the pole and one of the other tags could cause problems in some applications. For example, in an audio switching circuit (a preamplifier selector switch or something of this nature) the brief gap in continuity could produce a loud "click" from the loudspeaker and would obviously be undesirable.

The only component which has not been considered so far is the battery, and it was decided early on in the proceedings that a small 9 volt battery such as a PP3 was the most convenient power source for this application, and the circuit was designed to have a current consumption within the capabilities of this component. Just what is within the capabilities of a PP3 size battery is something which is not really clear cut since there are numerous manufacturers producing batteries of this size and it is unlikely that they all give exactly the same level of performance. There are also special versions of the PP3 for use in a particular application, such as a LED calculator.

Working from practical experience rather than from any battery data sheets I would recommend that a PP3 should not be used where a piece of equipment will be used for long periods of time at a current consumption of more than about 10 milliamps. The current consumption of the timer is typically a little under 10 milliamps when the alarm is not operating, but rises to perhaps 40 or 50 milliamps when the alarm is activated. This is acceptable because the unit will presumably be switched off soon after the alarm is activated and this level of current consumption will not be maintained for any length of time.

For applications where a supply current of around 10 to 50 milliamps is needed for long periods it is much better to use a larger battery such as a PP7 or PP9 type. In fact the PP9 is suitable for use at supply currents of up to about 100 milliamps. A PP3 size battery would not work well at currents of this order, and would almost certainly have a very limited operating life indeed. Also, as the internal resistance of a small 9 volt battery is relatively high, at currents of around 50 milliamps the output voltage of a fresh battery could well be substantially less than 9 volts.

It is unusual for battery operated equipment to need a supply current of much more than about 100 milliamps, but there are some items of equipment that do. In such cases it is normal to use a number of 1.5 volt cells in series, rather than a large 9 volt battery. Of course, by using four, six, or eight of these cells in series it is possible to obtain supply potentials of 6, 9 and 12 volts respectively, and you are not restricted to a 9 volt supply. Where bursts of very high current will be needed alkaline batteries (such as the MN1500 which is physically equivalent to an HP7 cell) are a good choice, or ordinary HP11 or HP2 cells could be used.

However, NiCad rechargeable cells are not all that expensive these days, and even allowing for the cost of a suitably battery charger these are likely to be a more economic choice in the medium and long terms. These cells have a very low internal resistance and there is little change in the output voltage as the cell discharges, until the cell becomes exhausted that is. Apart from the higher initial cost there is one slight drawback to these cells, and that is their output potential of only about 1.2 or 1.25 volts. This is obviously lower than an ordinary 1.5 volt cell, and is in fact comparable to a nearly exhausted 1.5 volt battery. In practice this shortfall is usually counteracted in part by the lower internal resistance, and NiCad cells can even give better performance than other types of cell in high current applications. However, in some cases it may be necessary to use one or two additional cells in order to compensate for the lower potential of each cell.

Before moving on to another project, refer back to the timer circuit of Figure 7 to see if you can understand the function of each component. If possible, build the circuit on a breadboard and experiment with a few of the values to see what effect changes have. If you understand the circuit properly you should be able to roughly predict what will happen if a particular change is made (e.g. halving the value of R7 would double the modulation frequency), and it can be very instructive to experiment in this way. Figure 10 gives a suitable component layout for this circuit using a Verobloc solderless breadboard.

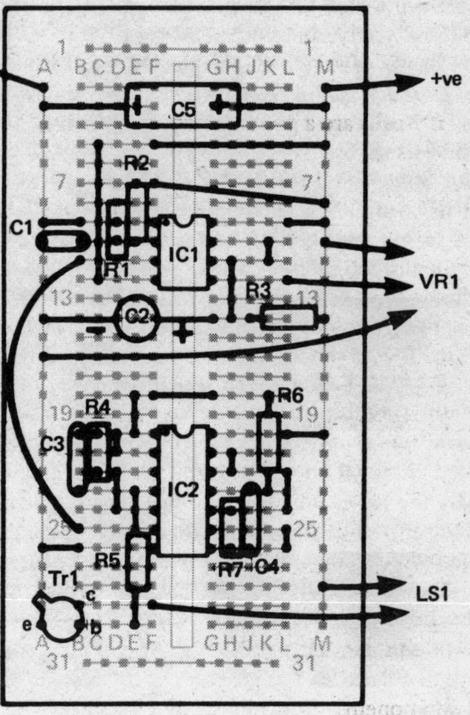

Fig. 10. Component layout for the Timer

Audio-Amplifiers

Audio amplifiers are still one of the most common types of circuit despite the increase in popularity of digital and other types of circuit. New types of electronic gadgets seem to appear constantly, but a sizeable proportion of these incorporate an audio amplifier of some kind or other.

Designing small and medium power audio output stages used to be an awkward task due to the use of Class B output stages which needed temperature compensation to prevent thermal runaway and the destruction of the output transistors due to overheating. Slightly too little thermal compensation and the circuit might still be damaged by overheating, while excessive compensation would result in an inadequate bias through the output transistors with severe cross-over distortion in consequenc

These days this problem does not arise since integrated circuit audio power amplifiers can be used. These generally give better performance and reliability than discrete designs, and are competitive in terms of cost. Unless you are particularly interest in the design of audio output stages there is not much point in using a discrete circuit.

There are a large number of audio power amplifier ICs available, and selecting a suitable device for a particular application could be a difficult task. However, there are just a few types which are popular and which are therefore widely available to the amateur user, and in most applications one of these more common devices will suffice. In general these devices are popular because they do the job well and are inexpensive, and there is no point in using one of the less common types unless this really does offer real advantages in your intended application.

LM380N

The LM380N is probably the best known and most popular audio amplifier currently in use. The reason for this popularity is almost certainly the low external parts count needed to produce a practical audio amplifier using this device, plus the creditable level of performance that is obtained. In its most basic form an LM380N audio amplifier requires just three discrete capacitors as shown in Figure 11. C1 is a DC blocking capacitor, and C3 provides supply decoupling.

The voltage gain of the LM380N is fixed at a nominal level of 34dB (50 times) by an internal negative feedback loop, and

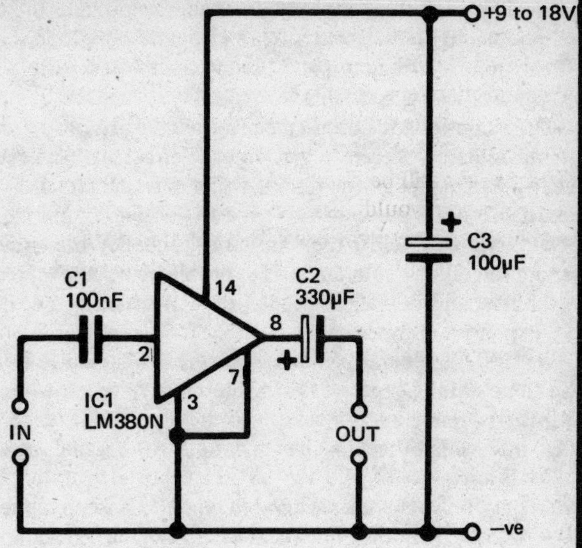

Fig. 11. Basic LM380N amplifier

while it is possible to boost or reduce the voltage gain of the circuit using discrete feedback components, this would not normally be necessary. A simple preamplifier can be added if more gain or a boost in input impedance is required, and an attenuator at the input can be used if a reduction in gain is required.

The LM380N can produce a respectable output power of over 500mW RMS into an 8 ohm impedance loudspeaker with a distortion level of only about 1% when using a 9 volt supply. This may be more than is needed in some applications, and a higher impedance loudspeaker could then be used. The maximum output power reduces roughly in proportion to any increase in the loudspeaker impedance. Using a 4 ohm impedance loudspeaker is permissible with the LM380N (if a suitable component can be obtained), and a 9 volt supply then gives an output power of more than 1 watt RMS. With a higher supply voltage the output power is boosted somewhat, and the device can give up to about 2 watts RMS into an 8 ohm load with an 18 volt supply, for example. The maximum permissible supply voltage is 22 volts incidentally.

The values of the coupling capacitors, in this or any other circuit, must be chosen to give an efficient signal transfer at the lowest frequency the circuit will handle. In an audio circuit it is not a good idea to have an efficient coupling at frequencies well below the minimum frequency the circuit will handle since this could encourage problems with low frequency instability, and in many cases it would mean using a quite large and expensive component.

20 Hertz is generally excepted as the lowest audio frequency, and the coupling capacitor must therefore have a low reactance at this frequency in comparison to the input impedance of the stage into which the capacitor is coupling the signal. A value of 2.2μF is adequate to give a minimal level of attenuation at the lowest audio frequencies when feeding into an impedance of 10 kilohms. For other impedances the value of the coupling capacitor would need to be changed, and the change in value must be inversely proportional to the variation in input impedance. For example, an input impedance of 100 kilohms is ten times higher than 10 kilohms and a coupling capacitor only one tenth of the value would be required, which is 0.22μF or 220nF in other words.

The input impedance of the LM380N is typically 150k, but can be substantially lower than this, and the circuit must be designed to meet worse case conditions (i.e. the value of C1 must be adequate for an LM380N which happens to have a fairly low input impedance, and not just for one which is a typical example or has a higher than usual input impedance). Designing for worse case conditions is something that always has to be kept in mind when deciding on circuit values.

Obviously the value of C1 is only adequate for an input impedance of about 220 kilohms or more, and the value of C2 is only adequate for a loudspeaker impedance of about 75 ohms or greater. Both of these components would therefore seem to be seriously under-valued, and if the circuit is to handle the full audio bandwidth they are. However, being realistic about it, there is little likelyhood that the circuit will need to handle low audio frequencies since it is almost certain to be used in conjunction with a fairly small loudspeaker mounted in a small enclosure, and this is likely to severely limit the bass response of the circuit. In fact it is unlikely that there would be any true bass output at all

Therefore, there is little point in using (say) a 470nF component for C1 and a 2200μF component for C2 in a practical situation. If the circuit is used in conjunction with a reasonably large loudspeaker in a decent sized enclosure then it would probably be worthwhile increasing C1 and C2 to perhaps 220nF and 1000μF respectively, but it is unlikely that a small amplifier of this type would ever be used with a really high quality loudspeaker.

The circuit of Figure 11 will be satisfactory for many applications, but sometimes the slightly more complicated circuit of Figure 12 will be needed. Here a volume control (VR1) has been added at the input, and there is no DC blocking capacitor between this and the input of IC1. The LM380N, in common with a number of other audio amplifier ICs, can be used with the input either biased to the negative supply rail or simply left floating. A coupling capacitor between VR1 and the input of IC1 would therefore be superfluous. A DC blocking capacitor is added ahead of VR1 so that any DC component on the input signal is blocked from the input of IC1 and is not able to upset the biasing of IC1. This also prevents any DC input component from being fed to VR1. This may not seem to be of any importance, but a DC signal across a volume control can have two detrimental effects. One is simply that as the slider is moved up and down the track a low frequency signal is generated, and this would probably be quite large in proportion to the input signal. In consequence, this signal could force the output of the amplifier fully positive or negative, making the circuit temporarily inoperative.

The second point is that with only a very small DC component present across the volume control the noise generated when the volume control is adjusted will be far greater and more noticeable than the noise level produced if there is no DC signal present.

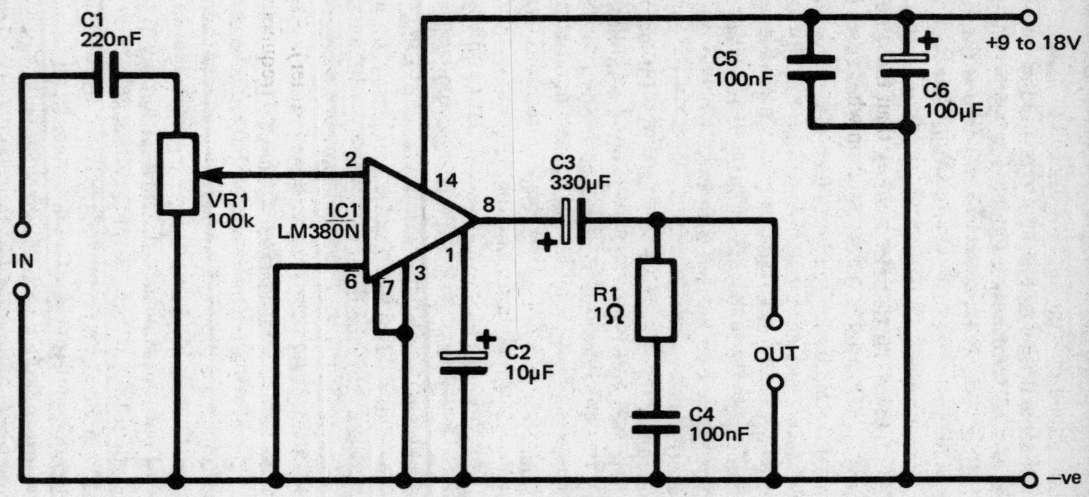

Fig. 12. A more comprehensive LM380N amplifier

This tends to be more noticeable when using a potentiometer which is becoming worn and "scratchy".

An interesting feature of the LM380N is that it has two input terminals; an inverting input at pin 6 and a non-inverting input at pin 2. In the circuits of Figures 11 and 12 it is the non-inverting input that is used, and in most applications it does not matter which one is used. In some situations it might be found that one input gives better stability than the other, and the LM380N can be used in a configuration which uses both inputs, but in most cases only one input will be used, and it will not matter which one. In Figure 12 the unused input is connected to the negative supply rail so that stray pick-up is avoided. This could possibly prevent instability or unwanted breakthrough at the output, but in most cases it will not be essential to do this.

C2 decouples the supply to the input stage of IC1, and this is useful if the circuit is powered from a simple mains power supply unit which has a fairly high level of ripple on the output. Without this capacitor the LM380N has a supply ripple rejection of only about 10dB. In other words, the hum level at the output of IC1 would only be about 10dB less than the hum level on the supply lines. Using a 10μF decoupling capacitor in the C2 position reduces the ripple level on the output to only about one hundredth of the supply ripple level (i.e. the ripple rejection is boosted to about 40dB). This should be sufficient to render the output hum level insignificant.

C5 is an additional supply decoupling capacitor, and this provides decoupling at high frequencies where C6 becomes inefficient (a technique that was mentioned earlier). Although the LM380N is an audio frequency circuit, the frequency response of the device extends well into the radio frequency spectrum. Although C5 will be unnecessary in many cases, it is probably worthwhile including it just in case. A decoupling capacitor of this type should be mounted physically as close as possible to the integrated circuit so that there is as little inductance and resistance as possible between the decoupling capacitor and the integrated circuit.

R1 and C4 are needed in order to aid stability, and a circuit of this type is sometimes called a "Zobel network". These are needed to provide a low impedance across the output of the amplifier at high frequencies where the highly inductive load provided by a loudspeaker can be very high. A network of this type is not always needed, and I have usually found it to be superfluous with the LM380N. Again, it might be worthwhile including these components just in case they are needed, although they could be easily added to the circuit by wiring them across the loudspeaker terminals or output socket, or the printed circuit could be designed to take them but the network would only be added if instability became evident. It is important that this type of network is added effectively in the leads going out to the loudspeaker or output socket, and not between (say) the output and the earth rail at the point where it is joined by the volume control. In other words, in the bus bar diagram of Figure 9(b) the line to the Zobel network would either come from the bus-bar at about the same point as the line to the loudspeaker, or it would be a branch from the line to the loudspeaker. If a Zobel network is wired into circuit carelessly it can easily cause instability rather than cure it!

The LM380N is suitable for many requirements, but it does have its limitations. It does not work well with a supply voltage of much less than about 9 volts for example, and the quiescent current can be as much as 25 milliamps. This makes the LM380N less than ideal for use in 9 volt battery operated equipment where only a modest output power is required, and it would almost certainly fail to operate at all using a nominal supply voltage of 3 or 6 volts. For an application of this type a device such as the TBA820M would be a better choice. This device operates well at supply voltages down to 3 volts, and the quiescent current consumption is typically only about 4 milliamps. The LM380N will provide a maximum output power of about 2.5 watts RMS, and obviously a somewhat higher output power will be needed in some applications. A higher power device such as a TDA2030 or an HA1388 would then be needed.

There is insufficient space available here to go into detail about several audio power amplifier devices, but data on the more popular types is not difficult to obtain, and it is usually quite easy to work out the function of each component if the available data is rather sketchy in this respect, and a suitable value can then be determined. Data sources for audio power amplifier devices normally give at least one circuit example with component values and a few basic performance figures, and it should be possible to adapt a circuit of this type to suit your requirements without too much trouble.

For further information the reader is recommended to see book number BP122: *Audio Amplifier Construction* by the same author and publisher as this book.

With most audio power amplifiers a few more discrete components will be required in comparison to the number used in the circuits of Figures 11 and 12. The LM380N, as pointed out earlier, can have either input floating or biased to the negative supply rail. With most other devices only one method or the other can be used, and the data or example circuit should make it clear which method must be used. The voltage gain of the LM380N is fixed at 34dB, but with most devices only one of the resistors in the negative feedback network is an internal component, and this permits the voltage gain of the circuit to be controlled using a discrete resistor. In most cases this resistor is wired between one pin of the IC and the earth supply rail via a DC blocking capacitor. The voltage gain of the amplifier is approximately equal to the value of the internal resistor divided by the value of the discrete resistor, and the data should specify the value of the internal resistor (this is nominally 6200 ohms for the TBA800 device for example).

The required closed loop voltage gain is simply equal to the RMS output voltage of the amplifier at full output divided by the RMS input voltage that is available. The RMS output voltage at maximum output is roughly one third of the supply voltage (this is accurate enough for our present requirements). Thus an amplifier which is powered from a 12 volt supply would give a maximum output of about 4 volts RMS, and if a nominal input signal level of 200 millivolts RMS was available the voltage gain of the amplifier would need to be 20 times (4V divided by 0.2V equals 20). The value of the internal feedback resistor divided by twenty would then give the required value for the discrete feedback resistor. In practice it would be necessary to use a resistor somewhat lower in value than the calculated figure to allow for component tolerances and the fact that the actual input signal could well be significantly lower than its nominal level. Thus, if the calculated value was 50 ohms, rather than choosing the nearest preferred value of 47 ohms, 39 ohms or even 33 ohms would be a more realistic choice.

The value of the capacitor in series with the discrete feedback resistor must be chosen to have a reactance value which is low in comparison to the feedback resistor's value at the lowest frequency the circuit will have to handle. In other words the value of this component is determined in the same way as for a coupling capacitor, but it is the value of the feedback resistor rather than input impedance which must be taken into account when choosing the value of this capacitor.

Some audio power ICs have bootstrapping, which is simply coupling some of the output signal back to the supply for the driver stage of the device so that a higher positive output voltage swing can be attained. This gives a slightly higher maximum output power for a given supply voltage. The bootstrapping is sometimes provided by a capacitor of around 100μF in value connected between the output terminal and "bootstrap" terminal of the device, but often a resistor from the "bootstrap" terminal to the positive supply is needed in addition. If maximising the output power is not necessary for your particular application, then it is quite in order to omit the bootstrapping capacitor. If the circuit has a bootstrapping resistor as a discrete component this must be retained in order to connect the positive supply rail through to the preamplifier, or it can be replaced by a shorting link.

Most of the more recent audio power amplifier ICs have built-in frequency compensation capacitors which help to prevent high frequency instability. Where a discrete component is required the relevant data sheet should give details of the effect various values have on the frequency response of the circuit together with recommended values, or at the very least there should be a circuit giving typical values which can be used.

Figure 13 shows a typical audio power amplifier configuration and gives the functions of the various components. Obviously not all IC audio power amplifiers use this configuration or something very similar. A few have both resistors of the negative feedback circuit as external components, and the DC blocking

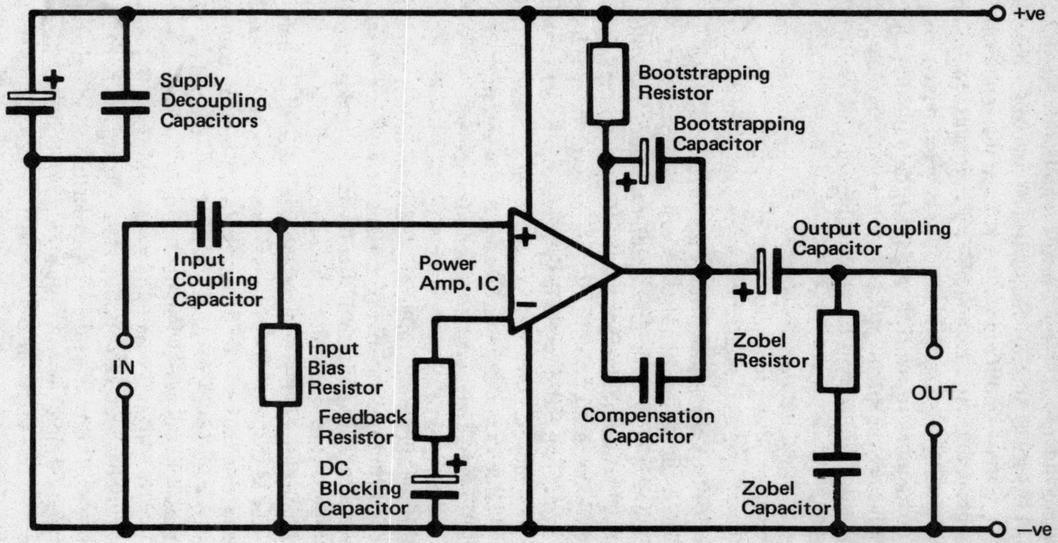

Fig. 13. A typical I.C. power amplifier configuration

capacitor may then connect between the junction of these two resistors and the inverting input of the amplifier (as in the case of the LM383 for example). Some devices are rather like high power operational amplifiers and apart from discrete feedback components also require external bias components (operational amplifier biasing and gain adjustment is covered next). By studying an applications circuit and other data it should be possible to discover the function of each component, and if necessary, how to choose a different value for your particular application.

Preamplifiers

When designing a preamplifier there are three basic types of circuit which can be used; discrete transistor, operational amplifier, and special audio preamplifier integrated circuits. The first option is one that has tended to gradually fall from favour as the integrated circuits for use in audio designs have steadily improved, but discrete designs are capable of very good performance indeed. However, from the design point of view integrated circuits are very much easier to use and probably represent the best starting point when producing your own designs.

For this reason we will only deal with operational amplifier and special audio preamplifier IC designs here. For general purpose use an operational amplifier (or two) provide great versatility, and it is very easy to produce circuits having the required input impedance and voltage gain figures. In their original role as DC amplifiers in analogue computers, operational amplifiers (which are so called because they perform mathematical operations) were used with dual balanced power supplies. They are often used in this way when used outside the analogue computing field, but in audio use there can be problems if dual balanced supplies and DC coupling are used, especially in a very high gain circuit. It is generally a lot easier to use a single supply even though this may increase the number of components required. We will therefore only consider single supply circuits here.

There are only two basic amplifying configurations in which

an operational amplifier can be used, the inverting mode and the non-inverting mode. These are shown in Figure 14 and Figure 15 respectively.

If we take the inverting circuit of Figure 14 first, the non-inverting (+) input of IC1 is biased to half the supply voltage by R3 and R4. A potential of half the supply voltage is chosen because the same voltage (within a few millivolts anyway) appears at the output of the amplifier, and this ensures that a large output voltage swing can be achieved before the output signal becomes clipped and seriously distorted.

Choosing values for R3 and R4 is not difficult since they must obviously be equal in value to give the required bias voltage, and the actual value chosen is not very critical. Less than a few kilohms would not be a good idea since this would unnecessarily increase the current consumption of the circuit, and a very high value would be inadvisable since it would then take a long time for C2 to charge up and for the circuit to settle down at its normal operating potentials. Also, with a non-FET input operational amplifier the input impedance might only be something in the region of one megohm, and this could result in the biasing being pulled away from its intended level by the shunting of IC1's input impedance on the biasing potential divider. Thus a value of anything from about 3k3 to about 100k should be satisfactory.

C2 is not essential, but it can have two beneficial effects. Firstly, any hum or noise on the supply lines which would otherwise be coupled by R3 and R4 to the non-inverting input of IC1 tends to be filtered out by R3 and C2 which act as a simple lowpass filter. An operational amplifier is largely unaffected by changes in the supply voltage, and quite large fluctuations in the supply voltage give only a barely detectable change in output potential. However, if this voltage change is allowed to find its way to one input of the operational amplifier it will be coupled straight through to the output, and in most cases it will appear substantially amplified at the output. The decoupling effect of C2 is therefore very important in many applications, but especially where a simple mains power supply is used as the power source and there is likely to be a fairly high level of ripple on the supply.

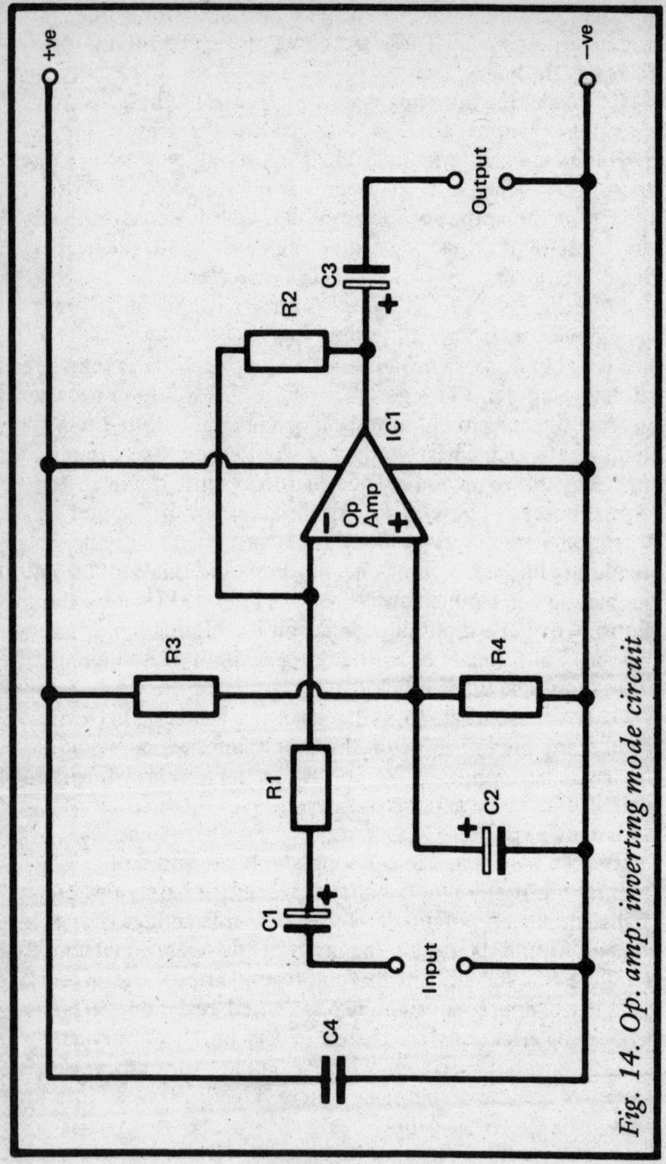

Fig. 14. Op. amp. inverting mode circuit

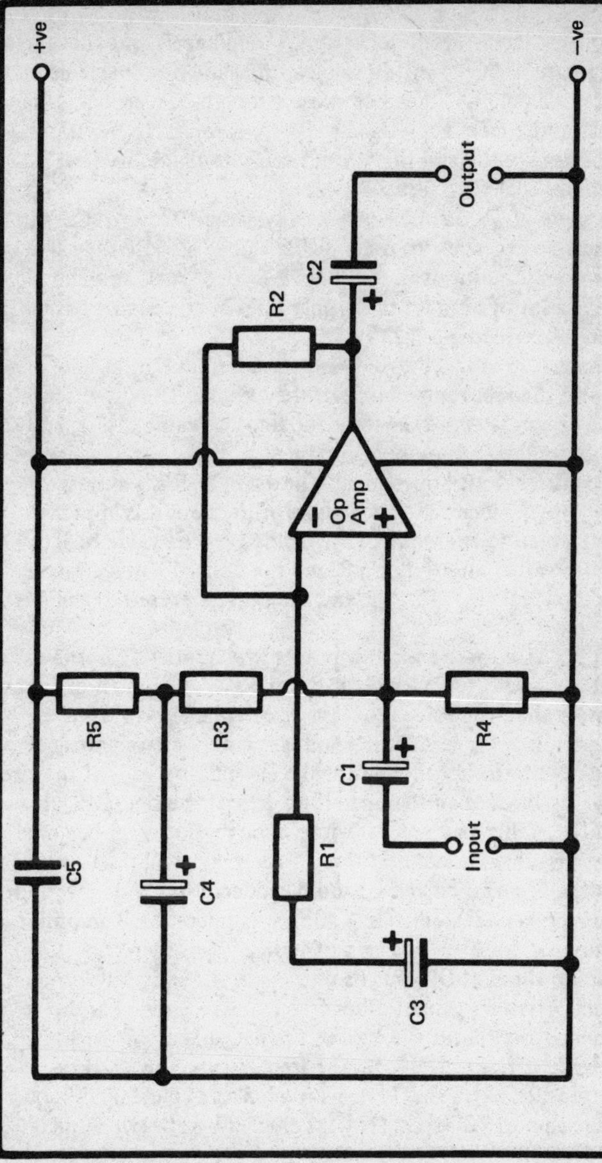

Fig. 15. Op amp non-inverting mode circuit

The second beneficial effect of C2 is that it provides a low impedance to the earth rail at audio frequencies and above. This greatly reduces any stray pick-up at the non-inverting input which might otherwise cause strong breakthrough of audio or even radio frequency signals. There is also a danger of stray feedback over the amplifier with resultant instability, but C2 eliminates this possibility.

A value of about 10μF is normally satisfactory for C2, but it might be necessary to use a higher value (or raise the value of R3 and R4) if a lot of decoupling is needed (such as when there is a lot of hum on the supply lines or the circuit has a high closed loop voltage gain).

The values assigned to R1 and R2 determine both the voltage gain and the input impedance of the circuit. The input impedance of the circuit is approximately equal to the value of R1, and so this component is simply given the nearest preferred value to the required input impedance. The value of R2 is almost as easily found, since the voltage gain of the circuit is approximately equal to the value of R2 divided by the value of R1. In other words, multiplying R1 by the desired voltage gain gives the ideal value for R2, and the nearest preferred value is then selected.

In practice there are a few points which must be borne in mind when selecting values for R1 and R2. The most important is simply that you cannot obtain more voltage gain from the amplifier as a whole than the voltage gain of the operational amplifier itself. In other words the closed loop gain of the circuit can be no more than the open loop gain of the operational amplifier. This may seem obvious enough, but what beginners sometimes overlook is that the voltage gain of an operational amplifier is specified at DC, and it reduces rapidly as the input frequency is increased. Thus, the 741C operational amplifier has an open loop voltage gain of typically about 100 or 200 thousand times at DC, but its unity gain – bandwidth product is just 1MHz. Its gain at other frequencies (where it is the frequency that limits the gain of the device) can be calculated simply by dividing 1MHz by the frequency involved. For example, at 50kHz the 741C would have a typical open loop voltage gain of 1000kHz (1MHz) divided by 50kHz, or just

20 times!

Obviously a single operational amplifier is not likely to give satisfactory results where high gain and the full 20kHz bandwidth is required, even given that most modern operational amplifiers offer a somewhat higher unity gain frequency (mostly about 3 or 4MHz). For high frequency use normal operational amplifiers are not really much use at all.

Another point to bear in mind is that there is a small input capacitance at the inverting input of the operational amplifier, and this can sometimes be significant, especially with FET input devices which generally have a higher input capacitance than bipolar types. One result of this capacitance is a reduction in the high frequency performance of the circuit if R1 is high in value, and it is also possible for a peak to appear in the response just below the start of the high frequency roll-off. There is also the problem that if high input impedance and high gain are required R2 would need to have in impractically high value. It is therefore advisable to only use this configuration where either high gain or high input impedance is needed, but not where both are required.

The value of C1 is chosen to suit the input impedance provided by R1, and the value of C2 should be selected to suit the input impedance of the stage fed from the output of the amplifier. In some cases this input impedance may be unknown, or the amplifier may be used with several pieces of equipment all having a different input impedance. The value of C3 is then chosen to suit the lowest input impedance with which the circuit will be used, or a value of 10µF is normally acceptable in cases when the input impedance of the following stage is unknown. C4 is simply a decoupling capacitor of about 100nF in value.

Non-Inverting Amplifier

With the non-inverting amplifier the non-inverting input is biased to about half the supply voltage, as for the inverting amplifier circuit, but it is obviously not possible to add a decoupling capacitor from this input to the earth rail as the input signal is coupled to the non-inverting input in this configuration. If decoupling is required, it must therefore be added in the positive supply connection to the biasing potential divider, and this is the purpose of R5 and C4. Provided the value of R5 is low in comparison to that of R3/4 the biasing will still have adequate accuracy if R3 and R4 have the same value, but remember that the upper bias resistor is effectively R3 plus R5 and that the value of R3 would therefore need to be reduced if R5 has a relatively high value.

The input impedance of the circuit is approximately equal to the parallel resistance of R3 and R4, and assuming that these two components have the same value, this value should be about double the required input impedance.

The voltage gain of the circuit can be calculated by first adding the values of R1 and R2, and then dividing this figure by the value of R1. Thus, if R1 had a value of 1k and R2 was a 10k component the voltage gain of the circuit would be eleven times (1k plus 10k equals 11k, 11k divided by 1k equals 11). In practice R1 would be given a reasonable value (anything between about 1k and 10k should be satisfactory), and the value required for R2 is then this value multiplied by one less than the required voltage gain. For example, a voltage gain of 23 times could be achieved with a value of 1k for R1, and R2 at one less than 23 times 1k, or 22 times 1k in other words which is obviously 22k. The value of C3 is chosen to have a reactance which is low in comparison to the value of R1 at the lowest frequency which the circuit must handle.

In theory the non-inverting mode circuit is capable of providing both high input impedance and high voltage gain, but in practice there is a likely snag if this is tried. Any stray feed-back from the output to the input of the amplifier will be in-phase and could easily stimulate instability. With the amplifier having high gain and a high input impedance only a very small amount of feedback would be needed to produce this instability and it would probably be impossible to obtain adequate isolation between the input and output of the circuit. A better solution is two use a two stage circuit which enables good isolation to be easily obtained, and where it is easy to design the circuit so that the input and output are out-of-phase and any feedback will be negative.

Special Preamp. ICs

Audio preamplifier integrated circuits are almost invariably basically operational amplifiers, but there is usually an internal bias circuit for the non-inverting input of the device, and in some cases there is also an internal feedback component or components. In fact these devices are very similar to audio power ICs as far as the basic circuit configurations are concerned, but they are designed for low noise rather than high output power.

The LM382 is a fairly typical example of an audio preamplifier IC, and Figure 16 shows the circuit of a simple preamplifier using one section of this device (it is a dual preamplifier device, like many other audio preamplifier ICs). C2 couples the input signal to the non-inverting input of IC1, and internal bias

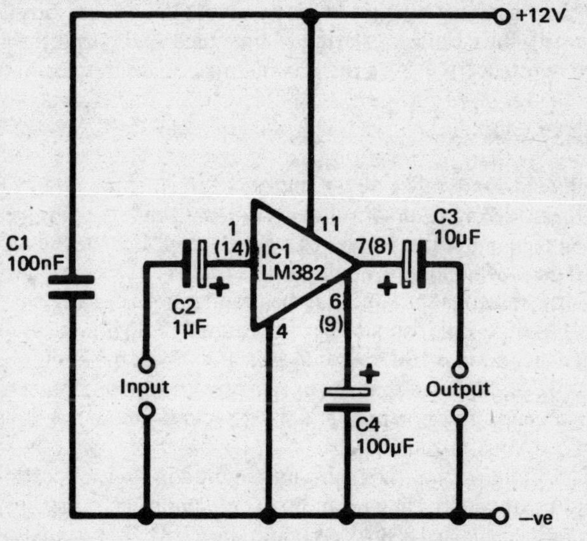

Fig. 16. An LM382 amplifier which gives 40dB of gain

components give a typical input impedance of about 150k. This could be boosted using a resistor in series with the input (with a consequent loss of gain) if necessary, or it could be reduced by adding a resistor across the input of the circuit. Having an internal bias circuit only presets the input impedance of the integrated circuit, not the input impedance of the circuit as a whole.

The LM382N has internal feedback components, and the only discrete feedback component required for a simple 40dB gain amplifier is DC blocking capacitor C4. The voltage gain can be modified by using different internal feedback components (which provides a very limited range of options) or by adding discrete feedback components. The latter can include capacitors to tailor the frequency response of the amplifier to suit applications that require equalisation (such as a tape or RIAA preamplifier). Again, having internal components may seem restrictive when compared to the flexibility which is possible using an operational amplifier or a discrete design, but this is not really the case, and it often enables an excellent preamplifier to be constructed using a minimal number of discrete components.

Bench Amplifier

Having discussed a few audio amplifier fundamentals we will now consider a few practical designs. As a starting point we will take an amplifier for use as a bench amplifier for the electronics workshop. An amplifier of this type should ideally be fairly sensitive so that it can be used with practically any signal source, and it should have a reasonably high input impedance so that it does not significantly load the circuit under test. We will assume that the unit is to be powered from a simple mains power supply and must provide an output power of about 2 watts RMS.

It is unlikely that the high gain required in this application could be obtained using an audio power amplifier alone and no preamplifier stage. A device such as the TBA820M can be set for a very high level of gain, but the problem with doing this is that the audio output quality is seriously degraded. This happens because the device is used practically at its open loop voltage gain, and the consequent lack of any negative feedback to counteract distortion gives a high level of distortion.

It is therefore necessary to use separate power amplifier and preamplifier stages, and the voltage gain provided by the power amplifier is not of great significance as the gain of the pre-amplifier can be set at a level which gives the required input sensitivity. The LM380N would seem to be a good choice for this application since it will give the required output power from a nominal 12 volt supply, it will work well using a simple mains power supply, and it requires few discrete components.

The LM380N requires an input of about 80mV RMS for maximum output when used with a 12 volt supply and the preamplifier must therefore have a voltage gain of about 80 times in order to give the required sensitivity of 1mV RMS. This is not a particularly high level of gain and can easily be achieved using a single operational amplifier, but things are complicated by the fact that a high input impedance is also required. It may therefore be necessary to use a two stage preamplifier with the first stage providing the circuit with a suitably high input impedance and the second stage giving the required voltage amplification.

However, an alternative that would be worth investigation would be to try having a non-inverting preamplifier with the required input impedance and voltage gain, followed by an LM380N used in the inverting mode. Although the preamplifier would, on its own, be likely to prove unstable due to stray feedback unless a great deal of care over the component layout was to be taken, the input and output of the amplifier as a whole would be out-of-phase. This would not guarantee freedom from instability, but it usually makes things a lot less critical and greatly reduces the possibility of problems arising.

So far we have not mentioned volume or tone controls, and the former would be essential in order to prevent strong signals overloading the amplifier, and when dealing with noisy signals a simple topcut tone control would also be a definite advantage. There are only two possible positions for the volume control; ahead of the preamplifier and between the preamplifier and power amplifier stages. As the preamplifier has a fairly high voltage gain it would be advisable to add the volume control before this stage rather than after it, as this avoids the possibility of a reasonably strong signal overloading the preamplifier. In fact an input level of only about 50mV RMS would be sufficient to overload the preamplifier.

A simple topcut tone control is not difficult to add to an audio amplifier, and circuits of this type normally just consist of a simple C – R lowpass filter with the resistance made variable so that the cutoff frequency can be varied. With the LM380N there is an interesting but equally simple alternative which can be used. This is shown in the skeleton circuit of Figure 17, and it consists of coupling the main signal to the inverting input and a filtered signal to the non-inverting input. In this case a highpass filter is used ahead of the non-inverting input so that only the treble frequencies are passed to this input. The effect of coupling signals to both inputs of the device is that one signal tends to cancel out the other so that no significant output is obtained.

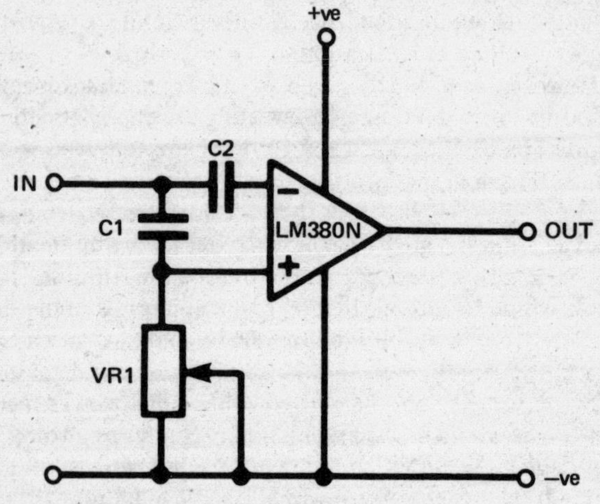

Fig. 17. Adding a tone control to the LM380N

If only a relatively weak signal is applied to the non-inverting input only partial cancelling occurs and the output signal level is reduced accordingly. As VR1 is adjusted for reduced resistance the cutoff frequency of the circuit is raised and the degree of treble cut is reduced. With VR1 at minimum resistance the non-inverting input is short circuited to the earth rail and there is no coupling to this input at all. The amplifier then has its normal, full bandwidth.

If the proposed amplifier was drawn out in block diagram form this would give something along the lines of Figure 18. With simple projects you may feel that a block diagram is unnecessary and that the overall make-up of the project can be envisaged without one. However, it is usually worthwhile quickly sketching out a block diagram because it will help to get things clear in your mind, and if there is some apparently minor point which is in fact crucial to the correct operation of the circuit it may help to bring this to light sooner rather than later. With medium and large projects a block diagram is really an essential first step in producing a finished design.

Turning the block diagram of Figure 18 into a final circuit should produce something along the lines shown in Figure 19. It must be emphasised that this circuit is just one solution to a problem which could be solved by an almost endless number of circuit permutations. It would be quite wrong to think that for a given application there is just one, definitive design.

If we now consider the circuit of Figure 19 in detail; IC1 is a straightforward non-inverting operational amplifier with R3 plus R4 biasing the non-inverting input and setting the input impedance of the amplifier at a nominal level of 500k. R2 and C2 provide hum filtering and supply decoupling for the bias circuit. Due to the high sensitivity of the circuit, plus the fact that it is designed for use with a simple mains power supply which will not have good regulation or a low ripple level, a substantial amount of smoothing is needed here. As R3 and R4 have a high value it is possible for R2 to have a fairly high value without affecting the biasing of IC1 too much. A value of 47k represents less than 5% of the value of R3 and is therefore reasonable. A value of about 10 to 22µF for C2 would, on the face of it, be quite adequate. However, when the circuit was

Fig. 18. Block diagram of the Bench Amplifier

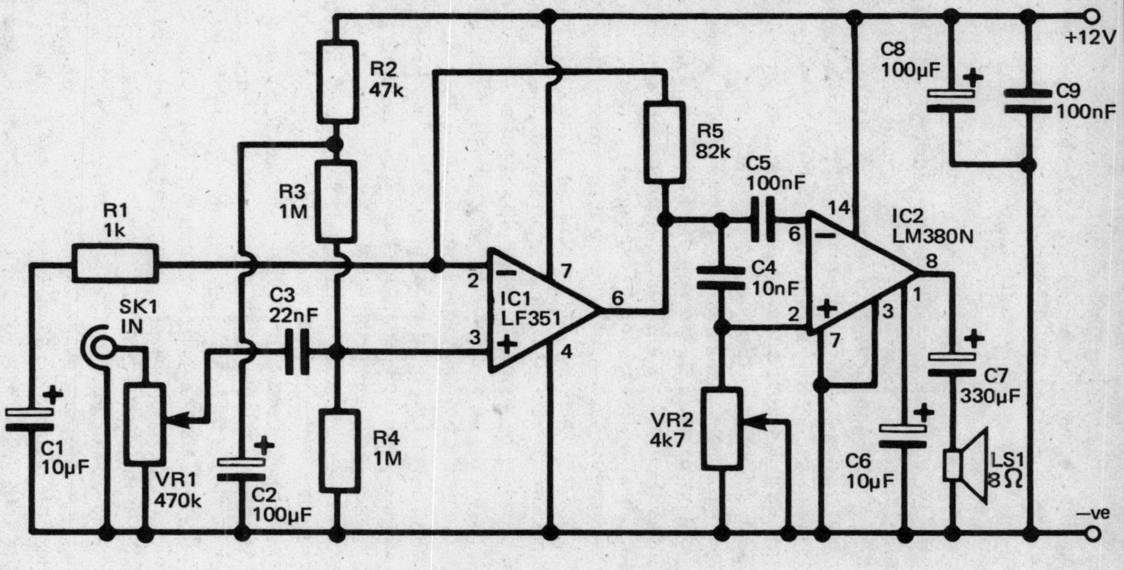

Fig. 19. The Bench Amplifier circuit diagram

tried it was found that a value somewhat higher than this was needed in order to give a really low hum level and to be sure of avoiding low frequency instability due to feedback through the supply lines. A value of 100μF is therefore specified as experience showed this to be necessary, rather than because the circuit was designed with this value.

R1 and R5 are the negative feedback resistors, and these obviously give a voltage gain of 83 times, which is close enough to the required figure of 80 times. C1 has a value which is chosen to suit R1 and the low frequency response required. C1 obviously gives somewhat less than a full bass response (which would require a value of about 2.2μF or more), but this is acceptable in this application where the loudspeaker used will not have an adequate low frequency response to do justice to the lower bass range. Similarly, the value of C3 is chosen to suit the input impedance of the preamplifier (1M plus 1M in parallel gives an input impedance of 500k), but an efficient coupling is not provided at the lowest audio frequencies.

Volume control VR1 should have a value which is roughly equal to or lower than the input impedance of the amplifier. In this case we require a fairly high input impedance and VR1 should therefore be no lower in value than is absolutely necessary since it shunts the input impedance of the amplifier. This gives an input impedance which varies from just under 250k at maximum volume (i.e. the parallel impedance of VR1 plus the input impedance of the preamplifier) to 470k at minimum volume (when the signal is only applied across the track of VR1). Note that a logarithmic potentiometer is used in volume control applications, and linear potentiometers are used in most other applications. This is due to the fact the human hearing has a logarithmic response and tends to provide a sort of compression so that changes in volume sound much less than they really are. Using a linear potentiometer as a volume control consequently gives unsatisfactory results with an abrupt increase in volume as the control is adjusted from the zero volume setting, followed by little or no apparent effect over the rest of the control's adjustment range. A logarithmic potentiometer provides little increase in volume over the initial part of its adjustment range followed by an abrupt increase in volume over the final part of its adjustment range, but to our hearing this gives what appears to be an even and regular increase in volume.

The power amplifier is quite straight forward, and as described earlier, as is the tone control. The values of C4 and VR2 are chosen to give a minus 6dB point (the point at which the voltage gain is about half its normal level) at a suitable frequency with VR2 at maximum resistance. In practice a minus 6dB point at about 3kHz would be the minimum usable, and the specified values give a cutoff frequency of approximately this figure. The minus 6dB point occurs at the frequency where the reactance of C4 is equal to the resistance of VR2. Reactances can be calculated using the appropriate formula, but reactance charts and tables are usually of adequate accuracy and are much quicker and easier in use. Figure 20 gives a brief but useful reactance (capacitive reactance that is) chart. VR2 has been given a value which is low in comparison to the input impedance of the LM380N so that the latter does not significantly shunt the value of VR2, and predictable results are obtained.

One final point concerns the component chosen for the IC1 position. A standard 741C operational amplifier is less than ideal for this application due to its gain-bandwidth product of 1MHz, and the fact that the preamplifier has a typical voltage gain of over 80 times. Multiplying the highest audio frequency of 20kHz by the gain of the amplifier (83 times) gives a figure of 1660kHz, or 1.66MHz in other words. Obviously the 741C cannot provide the required gain over the full audio range. Also, as the bias resistors have quite high values a FET input operational amplifier having an ultra-high input resistance would be advantageous.

The LF351 has a gain-bandwidth product of 4MHz, a FET input stage which gives an input impedance of typically 1 million megohms, a low noise level, and good distortion performance. It is therefore ideal for this application, but similar devices such as the TL081CP and TL071CP will work just as well in the circuit.

You might find it instructive to build this circuit and experiment with the circuit values. A suitable component layout using a Verobloc breadboard is given in Figure 21. If you do build the circuit you will notice that a combination of

F

C	20H	50H	100H	200H	500H	1k
100p			15.9M	7.59M	3.18M	1.59M
220p		14.5M	7.23M	3.61M	1.45M	723k
330p		9.64M	4.82M	2.41M	964k	482k
470p	16.9M	6.76M	3.38M	1.69M	676k	338k
680p	11.7M	4.68M	2.34M	1.17M	468k	234k
1n	7.59M	3.18M	1.59M	759k	318k	159k
2.2n	3.61M	1.45M	723k	361k	145k	72.3k
3.3n	2.41M	964k	482k	241k	96.4k	48.2k
4.7n	1.69M	676k	338k	169k	67.6k	33.8k
6.8n	1.7M	468k	234k	117k	46.8k	23.4k
10n	759k	318k	159k	75.9k	31.8k	15.9k
22n	361k	145k	72.3k	36.1k	14.5k	7.23k
33n	241k	96.4k	48.2k	24.1k	9.64k	4.82k
47n	169k	67.6k	33.8k	16.9k	6.76k	3.38k
68n	117k	46.8k	23.4k	11.7k	4.68k	2.34k
100n	75.9k	31.8k	15.9k	7.59k	3.18k	1.59k
220n	36.1k	14.5k	7.23k	3.61k	1.45k	723R
330n	24.1k	9.64k	4.82k	2.41k	964R	482R
470n	16.9k	6.76k	3.38k	1.69k	676R	338R
680n	11.7k	4.68k	2.34k	1.17k	468R	234R
1µ	7.59k	3.18k	1.59k	759R	318R	159R
2.2µ	3.61k	1.45k	723R	361R	145R	72.3R
3.3µ	2.41k	964R	482R	241R	96.4R	48.2R
4.7µ	1.69k	676R	338R	169R	67.6R	33.8R
6.8µ	1.17k	468R	234R	117R	46.8R	23.4R
10µ	759R	318R	159R	75.9R	31.8R	15.9R
22µ	361R	145R	72.3R	36.1R	14.5R	7.23R
33µ	241R	96.4R	48.2R	24.1R	9.64R	4.82R
47µ	169R	67.6R	33.8R	16.9R	6.76R	3.38R
68µ	117R	46.8R	23.4R	11.7R	4.68R	2.34R
100µ	75.9R	31.8R	15.9R	7.59R	3.18R	1.59R
220µ	36.1R	14.5R	7.23R	3.61R	1.45R	
330µ	24.1R	9.64R	4.82R	2.41R		
470µ	16.9R	6.76R	3.38R	1.69R		

Fig. 20. A useful A.F. capacitive reactance chart

2k	5k	10k	20k	50k
759k	318k	159k	75.9k	31.8k
361k	145k	72.3k	36.1k	14.5k
241k	96.4k	48.2k	24.1k	9.64k
169k	67.6k	33.8k	16.9k	6.76k
117k	46.8k	23.4k	11.7k	4.68k
75.9k	31.8k	15.9k	7.59k	3.18k
36.1k	14.5k	7.23k	3.61k	1.45k
24.1k	9.64k	4.82k	2.41k	964R
16.9k	6.76k	3.38k	1.69k	676R
11.7k	4.68k	2.34k	1.17k	468R
7.59k	3.18k	1.59k	759R	318R
3.61k	1.45k	723R	361R	145R
2.41k	964R	482R	241R	96.4R
1.69k	676R	338R	169R	67.6R
1.17k	468R	234R	117R	46.8R
759R	318R	159R	75.9R	31.8R
361R	145R	72.3R	36.1R	14.5R
241R	96.4R	48.2R	24.1R	9.64R
169R	67.6R	33.8R	16.9R	6.76R
117R	46.8R	23.4R	11.7R	4.68R
75.9R	31.8R	15.9R	7.59R	3.18R
36.1R	14.5R	7.23R	3.61R	1.45R
24.1R	9.64R	4.82R	2.41R	
16.9R	6.76R	3.38R	1.69R	
11.7R	4.68R	2.34R	1.17R	
7.59R	3.18R	1.59R		
3.61R	1.45R			
2.41R				
1.69R				
1.17R				

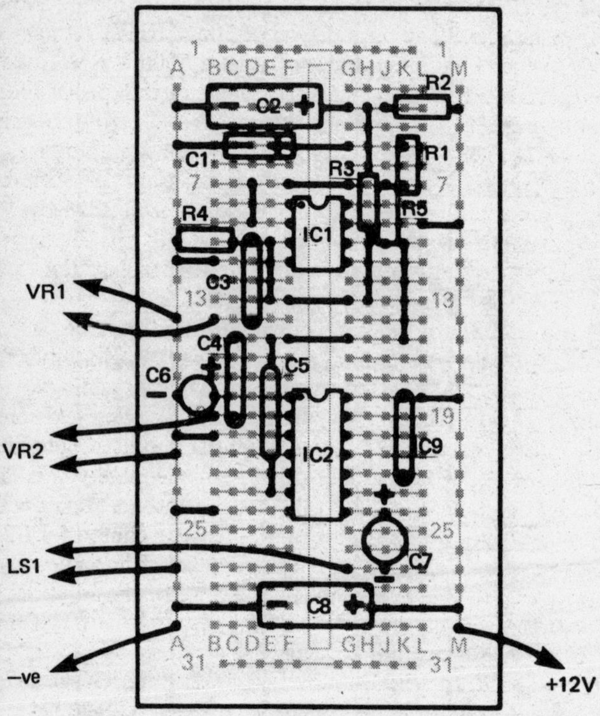

(N.B. pin 10 and pin 3 of IC2 are internally connected)

Fig. 21. Component layout for the Bench Amplifier

high input impedance and high sensitivity make all the wiring around the input of the circuit very sensitive to stray pick-up of mains hum and other electrical noise. This makes good screening of the input components and wiring very important, especially if a mains power supply is used.

Simple Preamplifier

If we now turn to a simple preamplifier application, let us assume that it is necessary to use an electric guitar having a low output pick-up (with an output level of 30mV RMS) with an amplifier which requires an input of at least 500mV RMS for full output. It would be possible to use one of the special audio preamplifier ICs in this application, but the level of voltage gain required is not very high and using an operational amplifier represents a lower cost but is perfectly satisfactory. We therefore require a preamplifier having a voltage gain of around 16.66 times and an input impedance of about 47k, and this can be achieved using a simple non-inverting amplifier such as the circuit shown in Figure 22.

R2 and R3 bias the non-inverting input of IC1, and as the unit will presumably be built as a self contained preamplifier powered from a 9 volt battery, there is no point in adding a hum filter in the positive supply feed to the bias circuit. There is no other stage in the design so that extensive supply decoupling is not required and C4 should be more than adequate. C2 is the input coupling capacitor, and in this application the circuit will need to have a good bass response and the value chosen for C2 is therefore more than adequate in this respect. The value of C3 can be chosen to suit the input impedance of the amplifier which will be fed from the output of the preamplifier but if this is unknown or the unit will be used with a variety of power amplifiers a value of 10μF should suffice. The chosen values for R2 and R3 give an input impedance of 50k incidentally, and this is obviously adequate for this application.

R1 and R4 set the nominal voltage gain of the circuit at 19 times, and even allowing for component tolerances this should give an adequate voltage gain, although a higher value could easily be substituted for R4 if experience proved a higher level of gain to be necessary. C1 has a value which maintains the bass response of the circuit well down to the lowest audio frequencies.

The choice of operational amplifier for this circuit is far less critical than in the previous example due to the lower voltage gain and input impedance of the preamplifier. The gain-bandwidth

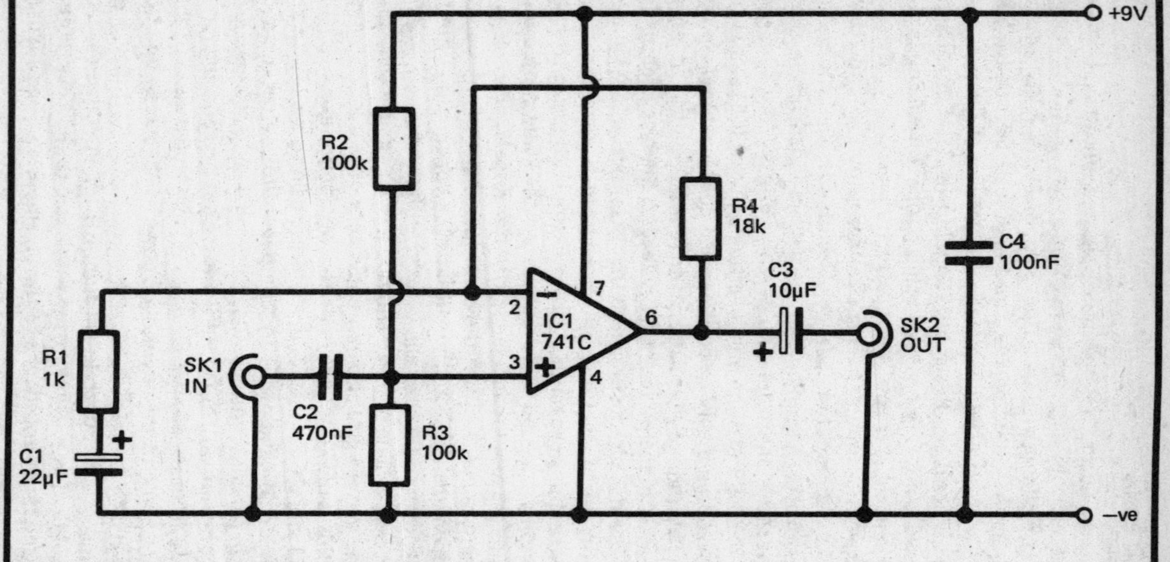

Fig. 22. The Guitar Preamplifier circuit

product of a 741C is more than adequate, since multiplying the voltage gain of the circuit by the highest audio frequency gives a figure of just under 400kHz. The input impedance, noise, and distortion performance of the 741C are all adequate for this application, and it is unlikely that any noticeable improvement in the audio quality would be produced if a higher quality audio operational amplifier such as a LF351 was to be used.

A suitable Verobloc layout for the amplifier is shown in Figure 23.

Simple Filtering

Where steep filtering is required it is necessary to use an active filter, and there are plenty of publications which deal in detail with this subject. However, sometimes simple 6dB per octave filtering is all that is required, or perhaps it is desirable to add simple filtering to an amplifier or mixer stage in a circuit to add to the attenuation provided by the main filter circuit. In either case it is extremely easy to add the filtering, and all that is required is one capacitor.

When using operational amplifiers (or operational amplifier style audio preamplifier ICs) the filter capacitor is added in the negative feedback circuit, and it is added in parallel with the resistor which connects between the output and inverting input of the operational amplifier. Figure 24 shows a filter capacitor added to a simple inverting mode circuit. This circuit is designed to have an input impedance of 10k, a gain of about 5 times, and a minus 6dB point at approximately 5kHz.

A circuit of this type is designed in the normal fashion, and then the appropriate value for the filter capacitor is selected. The minus 6dB point of the amplifier occurs at the frequency where the reactance of C4 is equal to the resistance of R4. In effect C4 is reducing the value of R4 at high frequencies and thus gives the required roll-off in the frequency response. Where the impedance of C4 is equal to the resistance of R4 the parallel impedance of the two components is half the value of R4, and the voltage gain of the circuit is half its normal level.

Reference to the reactance chart of Figure 20 shows that for a reactance of about 47k at 5kHz a capacitance of about 680pF

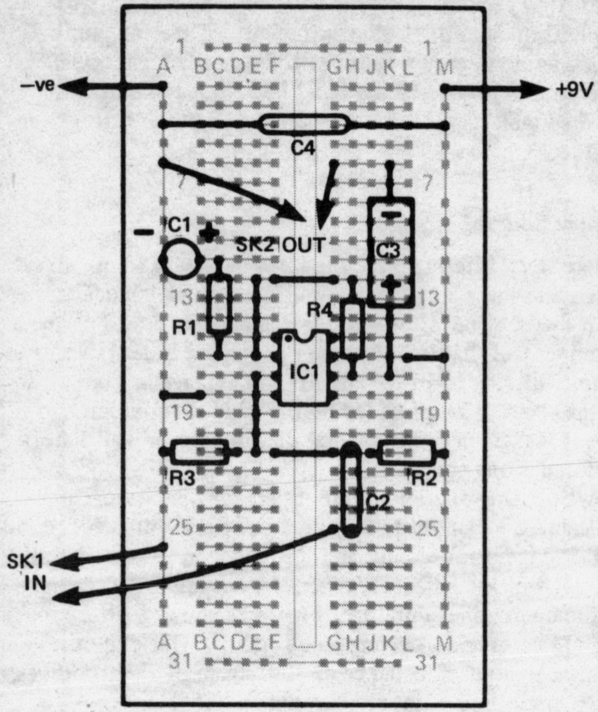

Fig. 23. Component layout for the Guitar Preamp

is required, and this is the value specified for C4.

This type of filtering does not work very well if used with a non-inverting amplifier, and in some cases would be completely useless. For example, if a non-inverting amplifier having a

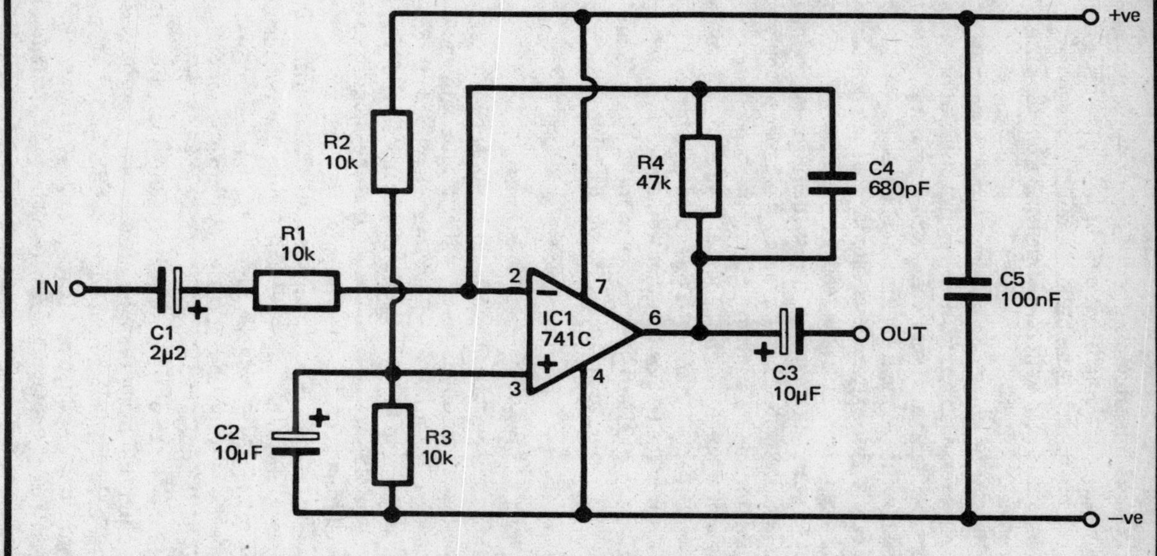

Fig. 24. Adding lowpass filtering to an inverting amplifier

voltage gain of two times was to have a filter capacitor added, the maximum roll-off applied to the circuit would be just 6dB (i.e. the gain of the amplifier would be halved). This is simply because a non-inverting amplifier cannot have a voltage gain of less than unity, even with zero impedance from the output to the inverting input.

In the case of a high gain non-inverting amplifier it is possible to obtain a reasonable amount of relative attenuation before the filtering reaches saturation level, but in general it is better to use the inverting mode if filtering is required.

Of course, it is only lowpass filtering that can be obtained by adding a capacitor across the feedback resistor. In theory it would be possible to obtain highpass filtering using an inductor instead of a capacitor, but in practice it is easier to apply simple highpass filtering by using a small coupling capacitor. For example, if we required the circuit of Figure 24 to have a minus 6dB point at about 200 Hertz, this could be achieved by using a capacitor in the C1 position that would have a reactance of about 10k at 200 Hertz. Reference to the chart of Figure 20 shows that a value of about 68nF would achieve this.

Microphone Preamplifier

As a final example of a simple audio amplifier circuit, let us assume that a microphone preamplifier for use with a low impedance (200 or 600 ohm) dynamic microphone is required, and that this amplifier must deliver an output level of (say) 200 millivolts RMS. We will also assume that the circuit is to be powered from a 9 volt battery supply.

The first step is to determine the main characteristics that the preamplifier must have. We will assume that the circuit is to be used as part of a public address system, and that as it will only be used for speech in a fairly noisy environment ultra low noise and the full audio range are not required. A response from about 50 Hertz to 5kHz is adequate for this application. The output from a low impedance dynamic microphone is typically around 200 microvolts (0.2 millivolts) RMS.

Thus the requirements are for a voltage gain of about 1000 times (60dB), an audio response from about 50 Hertz to 5kHz, and as a low impedance signal source is to be used the amplifier can have a fairly low input impedance. About 600 ohms is suitable for this applications.

The high voltage gain required rules out the use of a single operation amplifier, and it is necessary to use either two operational amplifiers or a high gain, low noise audio pre-amplifier IC. A twin operational amplifier circuit probably represents the most cost effective and simple approach to the problem, and we will therefore concentrate on a circuit of this type.

In the interest of obtaining good stability it is usually better to have a circuit where the input and output are out-of-phase so that any stray feedback over the amplifier is of the negative variety and is unlikely to cause instability. The only way to achieve this with a two stage circuit is to use one inverting stage and one non-inverting type, but it does not matter which we use at the input and which one is placed at the output. In practice it is probably slightly easier to use a circuit which has the inverting amplifier at the input as this permits a simple DC coupled circuit to be used, with the non-inverting amplifier being biased by the output of the inverting amplifier. Figure 25 shows a simple microphone amplifier which uses this technique.

The voltage gain should be roughly the same through each stage of the circuit so that niether of the amplifiers is operating at an excessively high gain figure. With the values shown IC1 has a closed loop gain of a little under 30 times while that of IC2 is nominally 40 times, and both amplifiers are therefore operating with reasonable levels of voltage gain.

R2 and R3 bias IC1, and although there is no mains hum on the supply lines which needs to be decoupled, it is advisable to include a capacitor in the C2 position to minimise any stray pick-up of noise and the possibility of oscillation due to stray feedback from the output of the circuit. R1 has the nearest preferred value to the desired input impedance of 600 ohms, and C1 has a value which gives adequate low frequency coupling. Note that although a minus 6dB point as high as 50 Hertz is acceptable, the value chosen for C1 gives a somewhat lower minus 6dB point. The amplifier as a whole has a minus 6dB

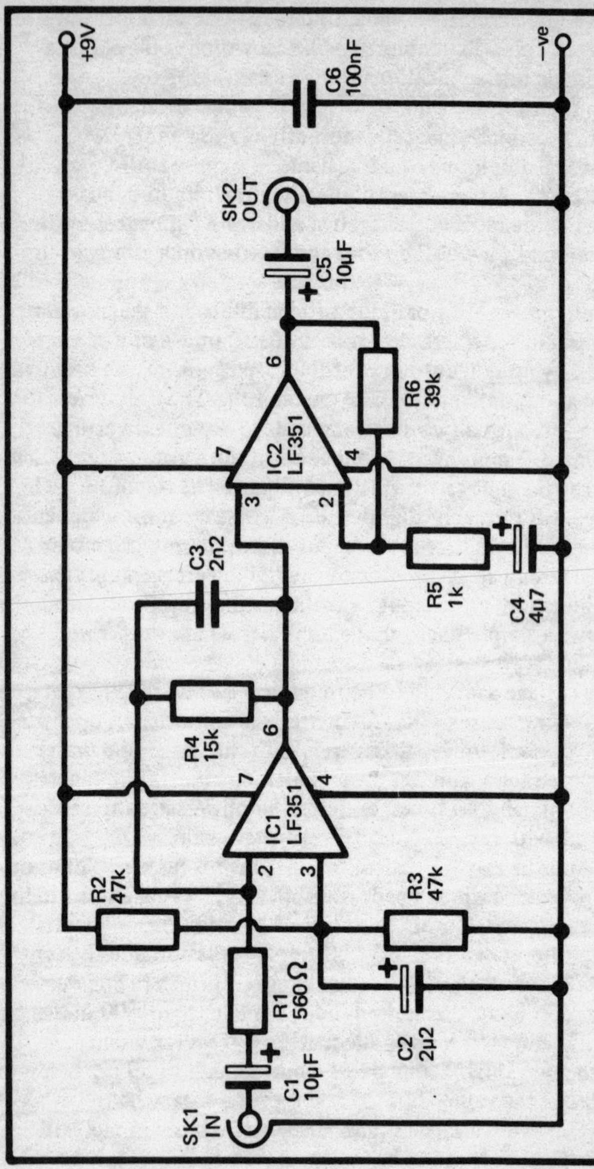

Fig. 25. A simple Microphone Preamplifier circuit

point at about 50 Hertz though, due to the additional low frequency roll-off provided by C4. However, it would be a perfectly valid way of doing things to select C1 to give the appropriate 50 Hertz cut-off frequency, and to then select the value of C4 to give no significant attenuation at 50 Hertz. Use whichever method you find easier.

In order to obtain an overall voltage gain of about 1000 times it is easy to see that each stage of the unit must give a voltage gain of about 30 or so, and R4 therefore needs to have a value in the region of 18k. For the reasons mentioned earlier, it is better to add the lowpass filter capacitor to the inverting amplifier rather than in the non-inverting one, and a capacitor of the appropriate value must therefore be added in parallel with R4. Looking at the reactance chart of Figure 20 a combination of a 15k resistor and a 2.2nF capacitor (which has a reactance of about 14.5k at the required 5kHz cut-off frequency) would seem to be the best values to use. This gives a slightly lower voltage gain than required, but the second stage can compensate for this without having to provide an excessive voltage gain.

The second stage is a straight forward non-inverting circuit, but as pointed out earlier, its input can be direct coupled to the output of the previous stage so that a bias circuit and coupling capacitor can be omitted. R5 and R6 give the circuit an overall gain which is somewhat higher than the desired figure of 1000 times, but this ensures that the circuit will have adequate gain even if the tolerances of the feedback resistors results in an actual voltage gain which falls slightly short of the calculated figure. A large excess of gain would be undesirable since this could conceivably result in the circuit driven from the output of the unit being overloaded, or more probably, it could result in the output stage of the preamplifier itself being overloaded. In this case the designed output level is about 200 millivolts RMS, which is roughly 560 millivolts peak to peak. IC2 will be able to produce a peak to peak output level of about three to four volts less than the supply voltage, or about ten times the normal output level in other words. This represents an overload margin of about 20dB which should normally be more than adequate.

At the levels of voltage gain involved here a normal 741C

operational amplifier would be operating near its gain limits, although the situation is eased somewhat by the fact that the full 20kHz audio bandwidth is not required. The circuit would probably operate quite well using two 741C devices, and if economy rather than performance was of prime importance it would be quite acceptable to use 741C operational amplifiers in place of the specified LF351s.

If you wish to experiment with this circuit the Verobloc component layout of Figure 26 can be used. You might like to try using different values of filter capacitor in the C3 position. You should notice that restricting the bandwidth of the circuit produces a very noticeable reduction in the noise level, and that the larger the amount of filtering used, the lower the noise level. Thus the use of filtering is not pointless and it does have certain advantages, one of which is this noise reduction.

Compressor

The design of straight forward amplifiers is a relatively simple task, and is one that is consequently a good starting point for beginners. However, in most cases you will probably want to design projects which are not entirely straight forward, and a simple example of an audio circuit of this type is a compressor. This is simply a circuit which does not have a fixed level of voltage gain, but instead has a level of voltage gain which reduces as the amplitude of the input signal is increased, so that the dynamic range of the processed signal is reduced. There is more than one type of compressor, and one type is used as a sort of limiter which prevents an excessive recording level when it is placed ahead of a tape recorder (in fact some recorders have such a circuit built-in). Circuits of this type are normally designed to have little or no effect until the input signal reaches a certain threshold level, and this level is just below the maximum acceptable recording level. The gain of the circuit then reduces rapidly as the input signal is taken above the threshold amplitude so that there is no significant increase in the output level.

As our next design example we will take a simple compressor for use as a recording limiter, with a nominal limiting level of (say) 500 millivolts RMS. In practice this limiting level will

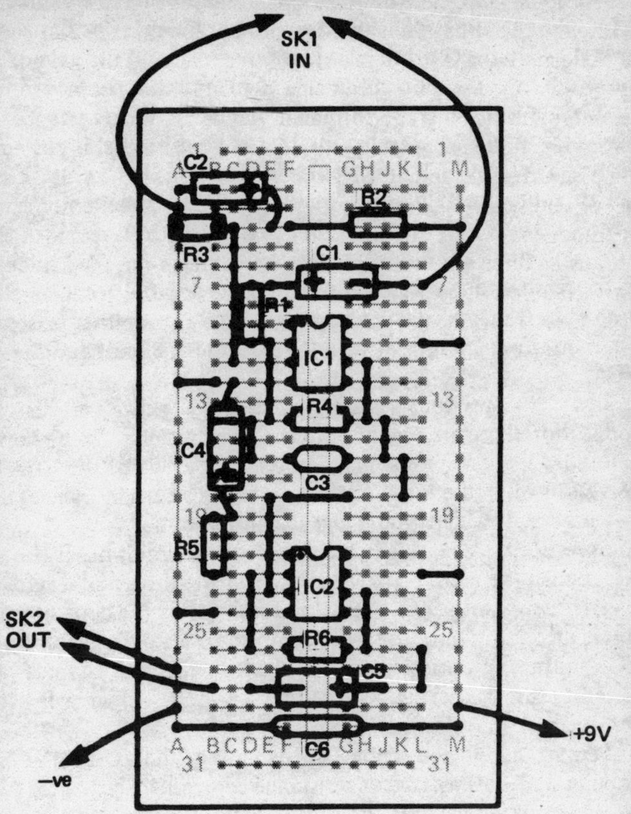

Fig. 26. Component layout for the Mic Preamp

actually need to be adjustable so that it can be trimmed to precisely the required level, but it must be designed so that the adjustment range is roughly centred around 500 millivolts RMS.

Any audio control circuit such as a compressor or expander is based on a voltage controlled amplifier, and must be designed so that a suitable control voltage is generated, and the gain of the circuit is controlled in the desired manner. In this case an arrangement of the type outlined in the block diagram of Figure 27 would seem to be suitable. Here the signal is passed through a VCA which under normal conditions has a voltage gain of approximately unity. However, some of the input signal is rectified and smoothed to produce a DC control signal for the VCA, and some additional amplification may be needed prior to the rectifier and smoothing circuit in order to drive it properly. It may also be necessary to have an amplifier or some other interface circuit between the smoothing circuit and the VCA.

In order to produce a circuit which will work well in this application there are two important points which must be borne in mind. Firstly, we must make sure that the gain of the VCA is decreased when the input signal goes above a certain level, and not increased (which would give expansion rather than compression). Also, it is important that the circuit has little effect until the input signal level exceeds the threshold level.

The first point is just a matter of interfacing the control voltage from the smoothing circuit to the control input of the VCA in the appropriate manner. Just how this is done must be varied to suit the particular type of VCA used, and we will consider a few possibilities here.

Perhaps the obvious choice as the basis of the VCA is the popular MC3340 voltage or resistance controlled attenuator chip. This is very simple to use, and little more than input and output coupling capacitors plus a compensation capacitor are required. In fact the only other component needed is a resistor to control the gain of the device, or alternatively a control voltage applied to this terminal. The basic circuit of an MC3340 VCA suitable for our present application is shown in Figure 28.

C2 is the compensation capacitor which aids the stability of the circuit by rolling off the frequency response above the audio range. A value of 620pF is recommended in the relevant data sheet, but a capacitor of this value would probably be

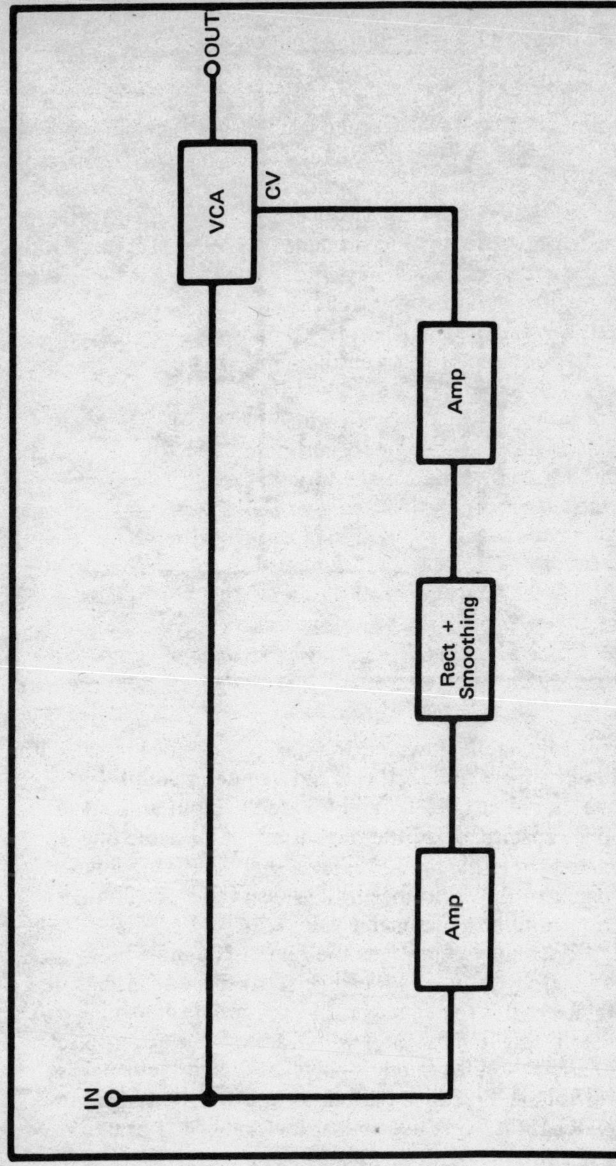

Fig. 27. A suitable arrangement for the Compressor

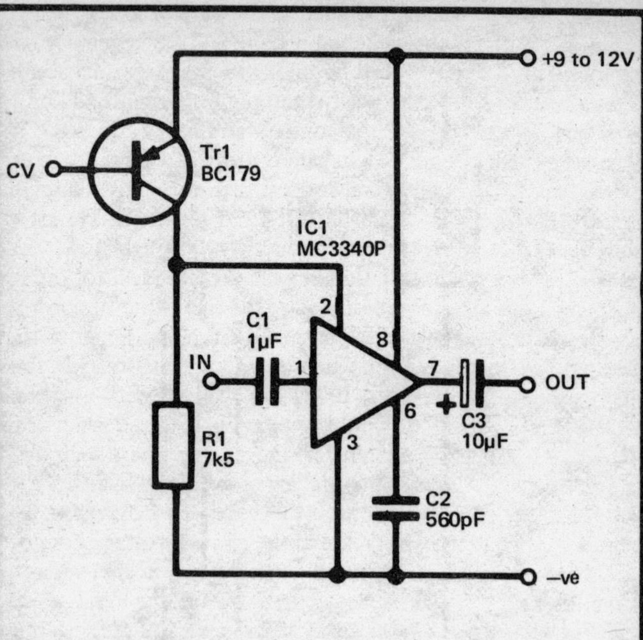

Fig. 28. A MC3340P VCA circuit

difficult to obtain. However, the precise value of this capacitor would not appear to be critical, and a value of 560pF (or 680pF) should suffice. C1 and C3 are the input and output coupling capacitors, and the value used for C1 is the one recommended in the MC3340P data sheet (with the input impedance of the device having a typical value of 50k this would seem to be a reasonable value).

Under stand-by conditions the gain of IC1 must be approximately unity, and it can be set at this level using either a suitable control resistance or the appropriate control voltage. In this application it would seem to be easier to set the quiescent voltage gain using a resistor, and then apply a control voltage to reduce the gain of the circuit at high input levels. R1 is therefore used to set the gain of IC1 at unity, and the control voltage is applied to the base of Tr1. The latter will normally be switched off and will consequently have no effect on the circuit, but by taking its base terminal about 0.6 volts or so below the positive supply potential it will be switched on and will pull the control terminal of IC1 more positive so that a reduction in gain is produced. The transfer characteristic of a silicon transistor is such that only a marginal increase above the base-emitter threshold voltage of Tr1 is sufficient to make the device conduct very strongly and produce a very large reduction in gain. This is ideal for our purposes since it automatically gives the right kind of control characteristic, and it is just a matter of arranging things so that Tr1 comes into conduction when the input signal reaches the maximum acceptable level. If Tr1 did not have such an obliging transfer characteristic it would be possible to add a zener diode in series with the base terminal so that it would not begin to conduct until the input potential exceeded the avalanche voltage of the zener. In fact it is quite easy to arrange the circuit so that a steady increase in the control voltage has no effect at first, followed by a sudden reduction in gain when some threshold level is exceeded, and it would be much more difficult to produce a circuit which gave a steady reduction in gain.

Note that Tr1 is a PNP device connected so that it pulls the control terminal of IC1 higher in voltage when it is biased into conduction, and not an NPN device connected across R1 so that it would give a lower control voltage. Reference to the MC3340P data sheet shows that a higher control voltage gives reduced gain, and the method of connection shown in Figure 28 is therefore the correct one. Reference to the MC3340P data sheet also gives a graph showing control resistance versus the relative gain of the device. It is important to note that it is the *relative* gain that the graph shows, and that (as stated in the data sheet) 0dB represents an actual gain of 13dB. The value of R1 must therefore be the one that gives a relative gain of –13dB, and this is about 7k5. A resistor of this value is available in the E24 series of values, and although in practice the gain of the circuit will not be precisely unity (due to component tolerances) it should be close enough for our purposes. However, R1 could be replaced with a 4k7 preset plus 4k7 fixed resistor in series so that the voltage gain could be adjusted to precisely the required level if this was considered to be important for some reason.

Of course, there are alternatives to using the MC3340P, and one of these alternatives is to use a VCA based on an operational transconductance amplifier such as the CA3080E or one section of the LM13600N. The normal configuration employed when a transconductance amplifier is used as a VCA is shown in Figure 29.

With this type of circuit the transconductance of the TOA (and hence the voltage gain of the circuit) is proportional to the bias current fed to the amplifier bias input. In this case R6 provides a bias current to this input which gives the circuit approximately unity voltage gain. Tr1 is connected so that it taps off some or all of this bias current when it is biased into conduction, and it therefore reduces the voltage gain of the circuit, very much like Tr1 of Figure 28.

Other forms of VCA circuit use opto-isolator arrangements, various types of field effect transistor, special ICs of various types, and a number of other arrangements. Which one is chosen for the final design depends on factors such as relative cost, noise performance, distortion performance, and the level of performance you require from the VCA. It is really just a matter of choosing the simplest or lowest cost option that will give at least adequate performance.

For this example we will base the compressor on the simple VCA circuit shown in Figure 30, and this has been taken from *Practical Electronic Building Blocks – Book 2*. This is a very crude and low-fi form of VCA, but it is suitable for our purposes in that it does not require any special components (virtually any operational amplifier can be used for IC1, and virtually any high gain NPN transistor will suffice for Tr1), and it is therefore easy for you to experiment with this circuit if you wish.

With Tr1 switched off the circuit functions as a simple inverting amplifier having its voltage gain set at roughly unity by R1, R3 and R6. If Tr1 is biased into conduction it forms a simple attenuator in conjunction with R1 and produces a reduction in the voltage gain of the circuit. Tr1 does not

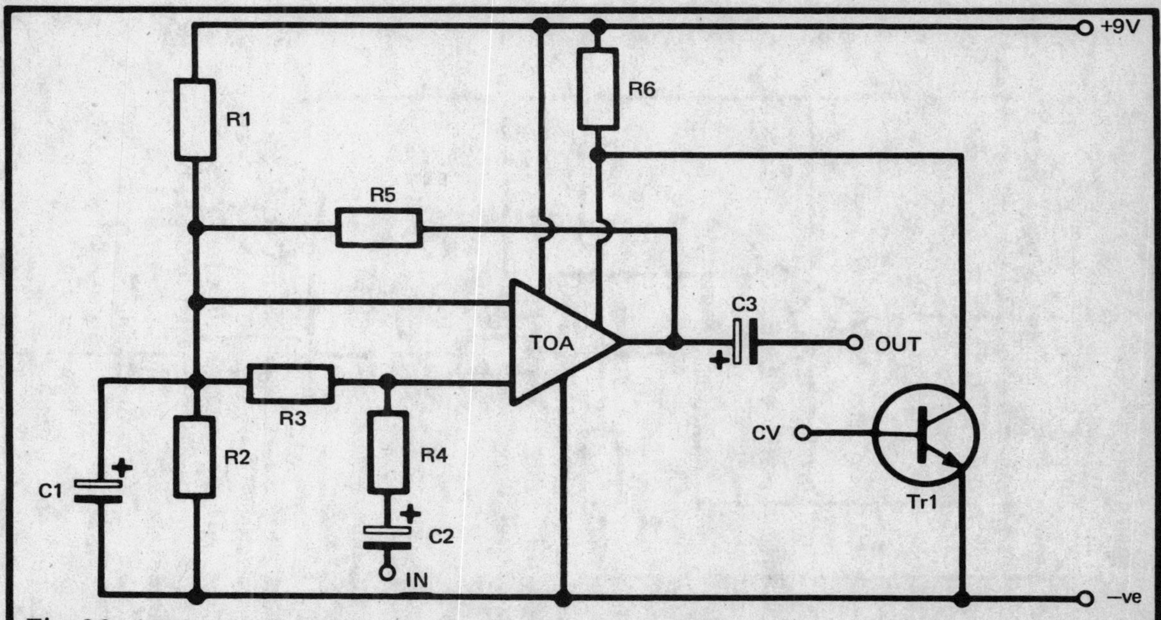

Fig. 29. A transconductance Op amp VCA configuration

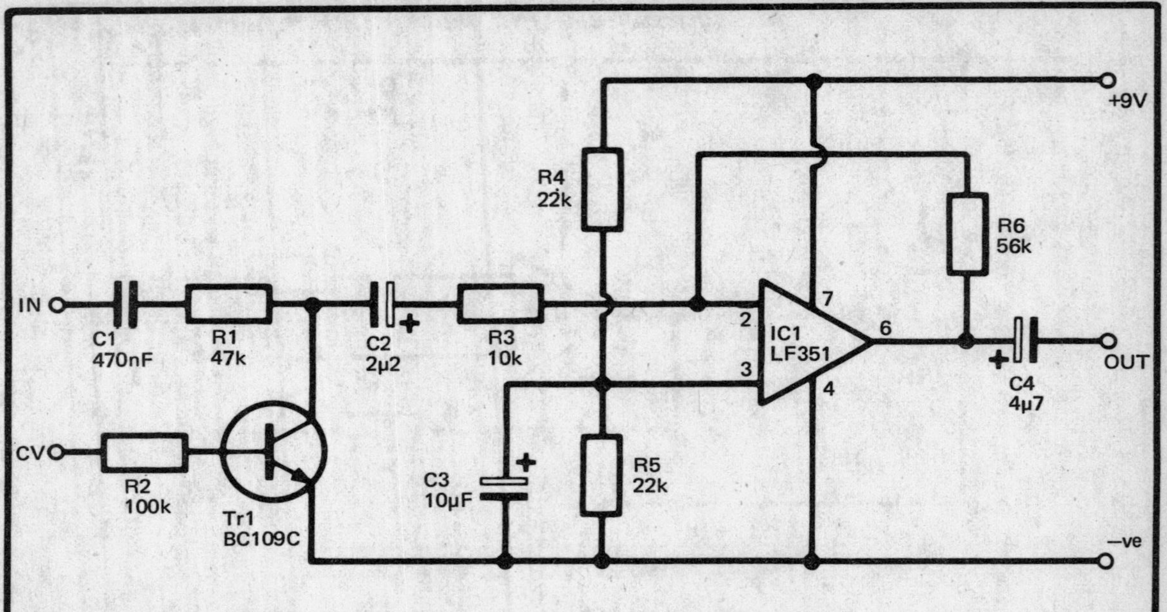

Fig. 30. A simple VCA circuit

provide a true resistance and therefore introduces a certain amount of distortion. However, results are adequate for many applications, and a simple circuit of this type is used in the automatic recording level system of many inexpensive cassette recorders.

The basic circuit has a fairly high value resistor in the R2 position so that an input voltage to the CV terminal of about 9 volts is needed in order to bias Tr1 into saturation and produce full attenuation. This gives a gradual introduction of the attenuation as the control voltage is increased, and for some applications this is desirable. For the present application though, a much sharper introduction of the attenuation would give better results, and a lower value for R2 would be beneficial.

Final Circuit

Choosing and adapting a suitable VCA is one thing, but using this in a final design is somewhat more difficult. However, it is not too difficult, and is made much easier if we refer back to the block diagram of Figure 27. We have the VCA block, and Tr1 is the amplifier block which feeds into the CV input of the VCA. In order to turn this basic VCA circuit into a compressor it is merely necessary to add a smoothing and rectifier circuit between the input and the control voltage input. An amplifier is likely to be needed ahead of the smoothing and rectifier circuit if the limiting is to commence at the desired level of 500 millivolts RMS since a signal at this level is unlikely to give an adequate control voltage. An amplifier would also act as a sort of buffer stage to minimise the loading on the input, and to avoid the non-linear loading of the rectifier and smoothing circuit on the input signal with the consequent distortion that would be produced. If the gain of the amplifier is made adjustable this would make it possible to set the input level at which the limiting commences at the desired level. A volume control style variable attenuator ahead of the amplifier would give basically the same effect. One final point is that the voltage gain of the amplifier does not need to be very high as a 500 millivolt RMS signal has a peak level of about 700 millivolts, and this is nearly enough to drive the CV input of the VCA (although there are losses through the rectifier circuit which must be taken into account). A voltage gain of only about three or four times should therefore be sufficient.

A simple circuit of the type shown in Figure 31 should therefore give the desired effect when added to the VCA of Figure 29. IC2 is used as a simple amplifier which can have its voltage gain varied from zero with R8 at minimum value to between four and five times with R8 at maximum value. R7 gives the circuit an input impedance of 47k, but this is shunted by the input impedance of the VCA to give the circuit as a whole an input impedance of around 25k. This should be satisfactory, but obviously a buffer amplifier could be added at the input if a higher input impedance was needed for some reason. The non-inverting input of IC2 is biased from R4 and R5 in the VCA circuit. A separate bias circuit could be used for IC2, but it should not be difficult to run a PCB track from the non-inverting input of IC1 to the non-inverting input of IC2, and this saves two resistors and a capacitor from the component count.

C6 couples the output of IC2 to a simple rectifier circuit which uses D1 and D2. The reason for including D1 in this circuit may not be immediately obvious, but it simply ensures that C6 is feeding into a reasonably symetrical load so that current can flow into and out of this component, and the required coupling will be obtained. If D1 is omitted it will only be possible for current to flow in one direction and no coupling will be obtained. A resistor of a few kilohms connected in place of D1 would achieve the same result, but C6 would then need to be made higher in value as it would be feeding into a lower impedance.

The load impedance into which C6 feeds is principally made up by R2 and the input impedance of Tr1. When D2 is biased into conduction it will effectively have a very low resistanc and will not add significantly to the load impedance. The input impedance of Tr1 is likely to be in the region of a few kilohms, but C6 can simply be given a value that will give good low frequency coupling even assuming an input impedance of zero at the base of Tr1. Note that R2 has been reduced by a factor of ten when compared to the original VCA circuit of Figure

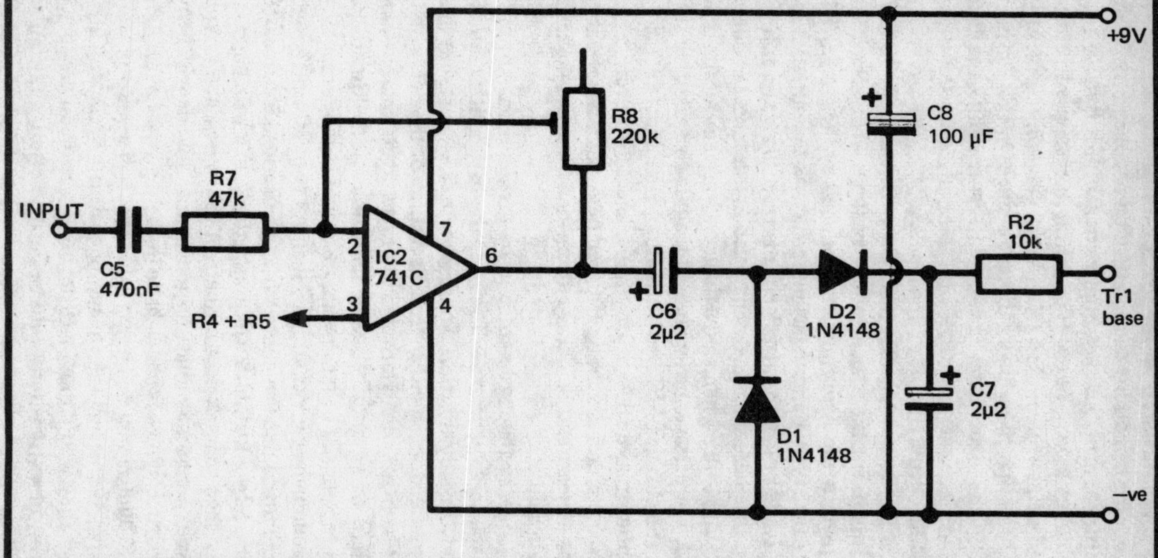

Fig. 31. Additional circuitry to give a Compressor action

30, and the reason for doing this was explained earlier.

C7 is the smoothing capacitor, and in applications of this type the value of this capacitor is crucial. If it is made too high in value the circuit will be too slow in its response to changes in input level, but if it is too small the circuit will respond too rapidly so that there are significant changes in gain during each cycle of the input signal. This would cause significant distortion and in an extreme case would give a totally distorted and unusable output.

It is important to realise that the circuit actually has two response times; the attack and decay times. The attack time is governed by the time constant produced by the output impedance of IC1 and the value of C7. As the output impedance of IC1 is very low the attack time is very short, and this is what we require here since the gain of the circuit must be reduced almost instantly when an overload occurs. The attack time could, if necessary, be increased by adding a resistor in series with the output of IC2, or decreased by adding a buffer amplifier to give a lower drive impedance. However, neither course of action should be required here and a reasonably fast attack time should be obtained provided C7 is not made very high in value.

The decay time is determined by the time constant of C7 plus the series resistance of R2 and the base-emitter junction of Tr1. Ideally this time would also be very short, but in order to avoid distortion either the attack time or the decay time must be made comparatively long to avoid continuous rapid changes in gain and the severe distortion referred to above. It is more important that the circuit should rapidly respond to (and eliminate) any overloads, than for it to return the gain of the circuit to its normal level after an overload has occured, and it is therefore logical to use a fast attack time with a much slower decay time. The decay time chosen has to be a compromise between a reasonably fast response time and one which gives an acceptably low level of distortion.

R2 and C7 could be regarded as rather like a lowpass filter which prevents audio frequency signals from reaching the VCA where they would modulate the input signal and produce distorti
Ideally an active filter would be used here so that a combination

of a fairly rapid decay time and high attenuation of audio frequencies could be attained. In a very simple circuit of the type envisaged here this would be somewhat out of place and over-complex, and a simple 6dB per octave filter with a suitably low cut off frequency has to be used.

From the reactance chart of Figure 20 it can be seen that the value chosen for C7 gives a reactance of about 3.6k at the lowest audio frequency of 20 Hertz, and this does not represent a great deal of attenuation at this frequency. However, the base-emitter resistance of Tr1 is effectively in series with R2 and helps to give greater attenuation, and a much higher level of attenuation is provided at higher freqencies where modulation of the input signal would be more noticeable. A value of around 2.2μF therefore represents a reasonable starting point, but with any critical component of this general type it is advisable to choose the final value by empirical means. Use the circuit in its intended set-up and try a higher value for C7 if modulation distortion is apparent, or a lower value (to see if a faster response time can be obtained) if it is not. Subjective assessment is the only way that the optimum value can be determined in a situation of this type.

Diodes

Diodes D1 and D2 do not have to pass particularly high currents since the 741C is simply not capable of providing more than about 20mA. With a supply voltage of only about 9 volts they do not need to withstand particularly high reverse voltages either, and any small signal diodes should suffice in this application.

There are two basic types of signal diode, germanium and silicon. Germanium devices have a much lower forward conduction threshold voltage and a forward resistance which steadily falls as the forward voltage is increased. These are useful in applications where small voltages are being processed and a minimal forward voltage drop is essential.

Silicon diodes require about 0.5 to 0.6 volts before they start to conduct significantly, and their forward resistance then falls sharply as the applied voltage is increased. In this application the higher forward voltage drop of silicon diodes is not really of any significance, and the greater forward voltage drop and sharper fall in forward resistance is actually beneficial in that it helps to give the compressor the required control characteristic. D1 and D2 can therefore be any general purpose silicon diodes such as 1N914s or 1N4148s.

Results

If you try out this circuit you should find that it does not seem to operate in quite the required manner. A suitable input signal can be taken from the earphone socket of a transistor radio or cassette recorder incidentally, and the output can be monitored using a crystal earphone. What you will probably find is that as the volume control of the radio or recorder is advanced, initially the volume rises, but as the limiting threshold is reached increasing the input level causes a reduction in the output signal level! Adjusting R8 will not improve matters either. If the voltage gain of IC2 is adequate the circuit will function in the manner described above, and if it is not the circuit will simply act as a buffer amplifier and have no real effect at all.

Hopefully you will be able to refer back to the block diagram of Figure 27 and discover the flaw in the design (if you have not already spotted it). If not, then I should perhaps explain that as the input signal is taken above the limiting threshold the control voltage to the VCA causes a large drop in the gain of the VCA. In fact the drop in gain more than counteracts any increase in the input signal level so that the output level actually falls.

While it would not be impossible to design a circuit where the control characteristic of the VCA was such that a reasonably consistent output level would be obtained with the input signal above the limiting threshold, this would be difficult as would setting up the circuit to operate properly. There is a much easier way of doing things, and you may be able to work this out for yourself (if you are not already familiar with automatic gain control techniques).

Really it is just a matter of rearranging the circuit slightly, as shown in the block diagram of Figure 32. This gives a sort

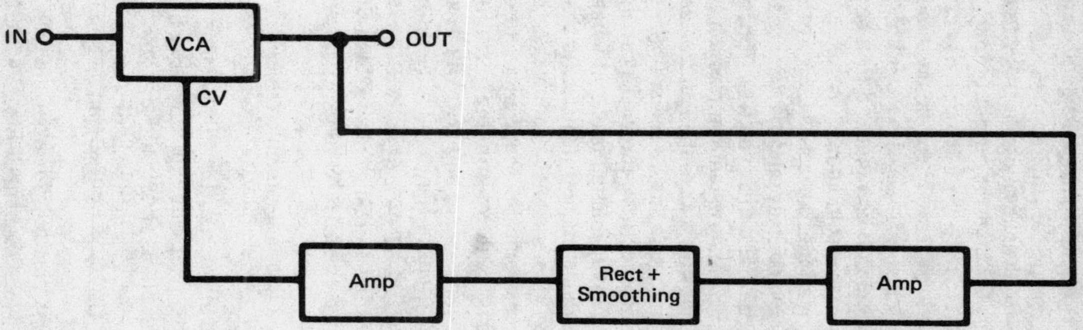

Fig. 32. Improved Compressor block diagram

of negative feedback loop from the output of the VCA to its control terminal, since a fall in output level due to an excessive control voltage would now produce a compensatory fall in control voltage.

Thus in this case the circuit can be made to operate properly simply by feeding C5 from the output of the VCA rather than the input. Unfortunately most design errors cannot be corrected this easily, and it is better to check things through and get them right first time if possible. However, things will inevitably go wrong from time to time and it will then be necessary to look carefully at the circuit, and the overall design, to work out just what is wrong so that corrective measures can be taken.

A suitable Verobloc component layout for the compressor is shown in Figure 33. This is for the final (functioning) version of the circuit. Figure 34 shows the compression graph for the prototype, and this shows a quite respectable level of performanc for such a simple circuit.

Sometimes when you have designed a circuit it is worthwhile having a final look at the design before building it up properly to ensure that nothing has been omitted, or that no superfluous components have been included. In this case there would appear to be a superfluous component in the form of C5. If the input of this amplifier is fed direct from the output of IC1 (rather than via C4) the input of the amplifier will be at the same potential as the output of IC1, or within a few millivolts of this potential anyway, and there is no potential difference for C5 to block. If you try out this circuit you should find that replacing C5 with a shorting link gives no detectable change in performanc

Final Points

In a publication of this size it is only possible to give a few simple design examples, but hopefully these illustrate some important design points and should give you the general idea of how to tackle the design of simple projects. Some of the more important points are summarized below:–

1. Learn as much as possible about modern electronic components and circuits. This is not something that can be

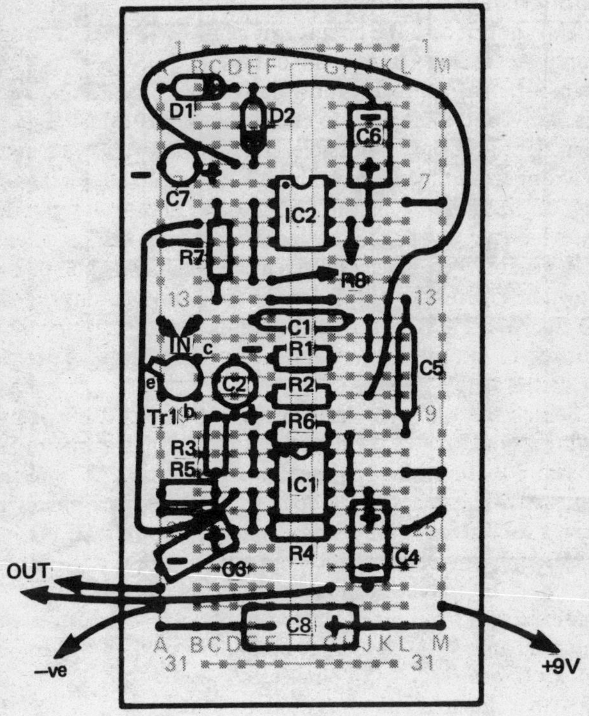

Fig. 33. Component layout for the Compressor

undertaken in a few days, or for that matter in a few months. However, with a certain amount of background theory you should be able to tackle a few simple designs, and as your knowledge of components and circuits grows it should be

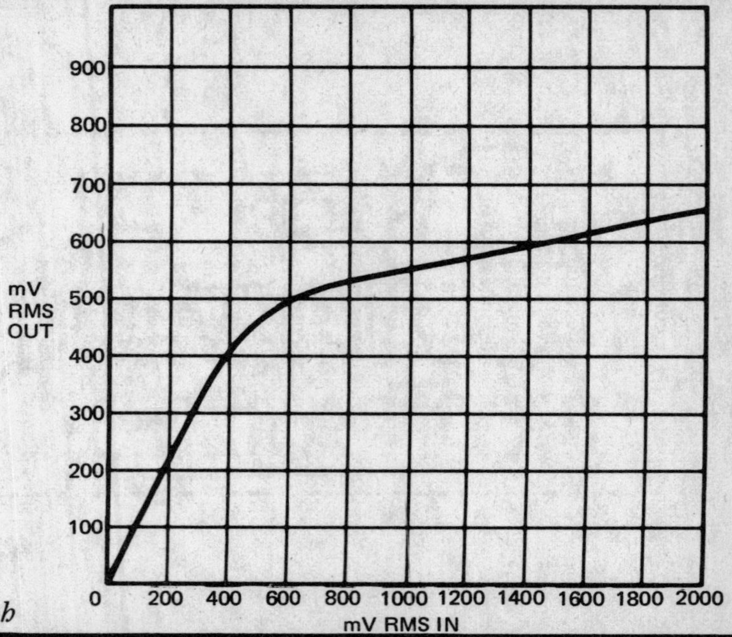

Fig. 34. Compressor performance graph

possible to progress to a few more complex projects without too much difficulty. With such a vast range of electronic components currently available and new ones appearing all the time this learning process is a continuous one. Sometimes you may not be able to find all the information you require, and it may then be necessary to experiment a little to see if a particular device or type of circuit can be made to operate properly in the particular application you have in mind. Try to go about things in a logical and well reasoned manner rather than simply trying every conceivable method of connection in the hope that something works!

2. A good background knowledge of electronics is really just the starting point these days. With so many circuits requiring the use of integrated circuits and other specialised semiconductor devices to make them feasable it is essential to have as much data on these as possible. It is essential to have at least basic data on devices of a type that is likely to be of use to you, but avoid wasting money on expensive data books which cover topics and devices which are unlikely to be of any use to you. Data books which cover areas such as discrete semiconductors, linear devices such as operational amplifiers and voltage regulators, and the standard TTL and CMOS families of logic ICs are likely to be of most use.

3. Be prepared to experiment and try out ideas. The fact that an integrated circuit or other circuit or device is not specifically intended for your particular application does not mean that it cannot be used in it successfully. Modern circuit design is largely a matter of using the available devices (particularly ICs) imaginatively. Using modern solderless breadboards it is relatively quick and simple to try out ideas, and even quite complex circuits can be built up in this way.

4. When you see a circuit design that interests you examine it in detail. Try to understand the function of each stage in the circuit, and preferably each component's function as well. You can learn a lot of useful circuit techniques in this way, and it can help to stimulate ideas of your own.

5. With the more simple projects you may find that you can simply make up a working prototype on a breadboard or Veroboard literally making up the unit as you go along, provided you have a good idea of the general arrangement of the unit instilled before you start. This is a perfectly valid way of doing things if it is a way that suits you. However, with more complex problems this approach is unlikely to be very satisfactory, and it is then virtually essential to work out a block diagram and at least a rough circuit before you start to put the first component onto the circuit board. An exception to this would be where you are unsure about the suitability of a device or stage of the circuit, and (if possible) it is then quite in order to test this stage independently. Indeed, it would be prudent to do so.

6. Inevitably there will sometimes be projects which prove to be impractical either because there is part of the circuit which proves to be beyond your capabilities, or because the cost and complexity of the circuit would make it impractical. In such cases it is a good idea to keep the project in mind when looking at new devices (or devices which are new to you). Sooner or later you will probably come across something that suits your requirements and enables the project to be completed successfully.

7. When you have worked out a basic circuit check through it carefully for design flaws. Remember to design for worst case conditions taking into account factors such as the minimum gain of the transistor type used, maximum tolerances of resistors, capacitors, and similar components, and so on. Make sure that the components are all used within their limitations with no resistors or semiconductors operating at excessive power levels, capacitors operating at excessive voltages, semiconductors operating at excessive voltages or currents (or combinations of these), or anything of this nature. Switches are about the most simple components imaginable, but it is remarkably easy to make a careless mistake when designing even fairly simple switching circuits. Check switching arrangements thoroughly making quite sure that none of the components are in danger of destruction at certain combinations of switch settings, and that short circuits on the supply lines or something of this type cannot be produced.

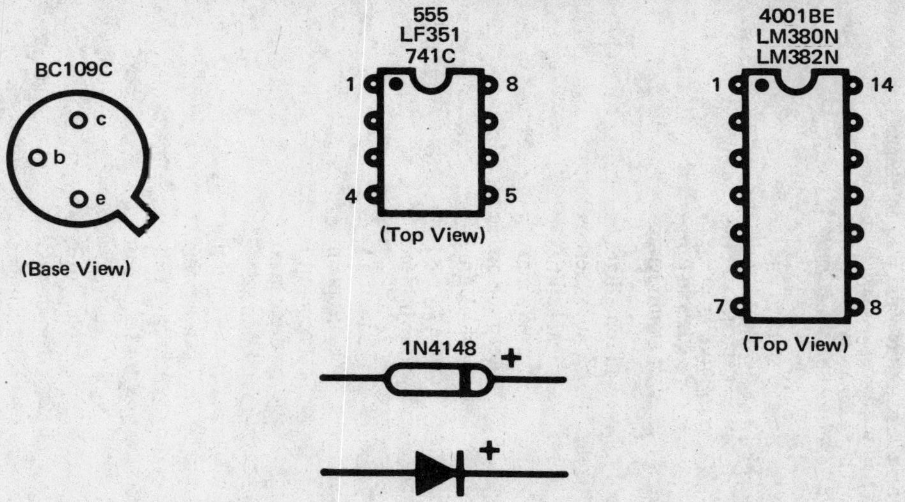

Fig. 35. Semiconductor leadout and pinout details

Please note following is a list of other titles that are available in our range of Radio, Electronics and Computer Books.

These should be available from all good Booksellers, Radio Component Dealers and Mail Order Companies.

However, should you experience difficulty in obtaining any title in your area, then please write directly to the publisher enclosing payment to cover the cost of the book plus adequate postage.

If you would like a complete catalogue of our entire range of Radio, Electronics and Computer Books then please send a Stamped Addressed Envelope to:

BERNARD BABANI (publishing) LTD
THE GRAMPIANS
SHEPHERDS BUSH ROAD
LONDON W6 7NF
ENGLAND

160	Coil Design and Construction Manual	£1.95
202	Handbook of Integrated Circuits (IC's) Equivalents & Substitutes	£1.95
205	First Book of Hi-Fi Loudspeaker Enclosures	£0.95
208	Practical Stereo and Quadrophony Handbook	£0.75
214	Audio Enthusiasts Handbook	£0.85
219	Solid State Novelty Projects	£0.85
220	Build Your Own Solid State Hi-FI and Audio Accessories	£0.85
221	28 Tested Transistor Projects	£1.25
222	Solid State Short Wave Receivers for Beginners	£1.95
223	50 Projects Using IC CA3130	£1.25
224	50 CMOS IC Projects	£1.35
225	A Practical Introduction to Digital IC's	£1.75
226	How to Build Advanced Short Wave Receivers	£1.95
227	Beginners Guide to Building Electronic Projects	£1.95
228	Essential Theory for the Electronics Hobbyist	£1.95
RCC	Resistor Colour Code Disc	£0.20
BP1	First Book of Transistor Equivalents and Substitutes	£1.50
BP2	Handbook of Radio, TV, Ind & Transmitting Tube & Valve Equivalents	£0.60
BP6	Engineers and Machinists Reference Tables	£0.75
BP7	Radio and Electronic Colour Codes and Data Chart	£0.40
BP14	Second Book of Transistor Equivalents & Substitutes	£1.75
BP24	52 Projects Using IC741	£1.75
BP27	Chart of Radio, Electronic, Semi-conductor and Logic Symbols	£0.50
BP28	Resistor Selection Handbook	£0.60
BP29	Major Solid State Audio Hi-Fi Construction Projects	£0.85
BP32	How to Build Your Own Metal and Treasure Locators	£1.95
BP33	Electronic Calculator Users Handbook	£1.50
BP34	Practical Repair and Renovation of Colour TVs	£1.95
BP36	50 Circuits Using Germanium, Silicon and Zener Diodes	£1.50
BP37	50 Projects Using Relays, SCR's and TRIACS	£1.95
BP39	50 (FET) Field Effect Transistor Projects	£1.75
BP42	50 Simple L.E.D. Circuits	£1.50
BP43	How to Make Walkie-Talkies	£1.95
BP44	IC 555 Projects	£1.95
BP45	Projects in Opto-Electronics	£1.95
BP47	Mobile Discotheque Handbook	£1.35
BP48	Electronic Projects for Beginners	£1.95
BP49	Popular Electronic Projects	£1.95
BP50	IC LM3900 Projects	£1.35
BP51	Electronic Music and Creative Tape Recording	£1.95
BP52	Long Distance Television Reception (TV-DX) for the Enthusiast	£1.95
BP53	Practical Electronics Calculations and Formulae	£2.95
BP54	Your Electronic Calculator and Your Money	£1.35
BP55	Radio Stations Guide	£1.75
BP56	Electronic Security Devices	£1.95
BP57	How to Build Your Own Solid State Oscilloscope	£1.95
BP58	50 Circuits Using 7400 Series IC's	£1.75
BP59	Second Book of CMOS IC Projects	£1.95
BP60	Practical Construction of Pre-amps, Tone Controls, Filters & Attenuators	£1.95
BP61	Beginners Guide To Digital Techniques	£0.95
BP62	Elements of Electronics – Book 1	£2.95
BP63	Elements of Electronics – Book 2	£2.25
BP64	Elements of Electronics – Book 3	£2.25
BP65	Single IC Projects	£1.50
BP66	Beginners Guide to Microprocessors and Computing	£1.95
BP67	Counter, Driver and Numeral Display Projects	£1.75
BP68	Choosing and Using Your Hi-Fi	£1.65
BP69	Electronic Games	£1.75
BP70	Transistor Radio Fault-Finding Chart	£0.50
BP71	Electronic Household Projects	£1.75
BP72	A Microprocessor Primer	£1.75
BP73	Remote Control Projects	£1.95
BP74	Electronic Music Projects	£1.75
BP75	Electronic Test Equipment Construction	£1.75
BP76	Power Supply Projects	£1.95
BP77	Elements of Electronics – Book 4	£2.95

BP78	Practical Computer Experiments	£1.75
BP79	Radio Control for Beginners	£1.75
BP80	Popular Electronic Circuits – Book 1	£1.95
BP81	Electronic Synthesiser Projects	£1.75
BP82	Electronic Projects Using Solar Cells	£1.95
BP83	VMOS Projects	£1.95
BP84	Digital IC Projects	£1.95
BP85	International Transistor Equivalents Guide	£2.95
BP86	An Introduction to BASIC Programming Techniques	£1.95
BP87	Simple L.E.D. Circuits – Book 2	£1.35
BP88	How to Use Op-Amps	£2.25
BP89	Elements of Electronics – Book 5	£2.95
BP90	Audio Projects	£1.95
BP91	An Introduction to Radio DXing	£1.95
BP92	Easy Electronics – Crystal Set Construction	£1.75
BP93	Electronic Timer Projects	£1.95
BP94	Electronic Projects for Cars and Boats	£1.95
BP95	Model Railway Projects	£1.95
BP96	C B Projects	£1.95
BP97	IC Projects for Beginners	£1.95
BP98	Popular Electronic Circuits – Book 2	£2.25
BP99	Mini-Matrix Board Projects	£1.95
BP100	An Introduction to Video	£1.95
BP101	How to Identify Unmarked IC's	£0.65
BP102	The 6809 Companion	£1.95
BP103	Multi-Circuit Board Projects	£1.95
BP104	Electronic Science Projects	£2.25
BP105	Aerial Projects	£1.95
BP106	Modern Op-Amp Projects	£1.95
BP107	30 Solderless Breadboard Projects – Book 1	£2.25
BP108	International Diode Equivalents Guide	£2.25
BP109	The Art of Programming the 1K ZX81	£1.95
BP110	How to Get Your Electronic Projects Working	£1.95
BP111	Elements of Electronics – Book 6	£3.50
BP112	A Z-80 Workshop Manual	£2.75
BP113	30 Solderless Breadboard Projects – Book 2	£2.25
BP114	The Art of Programming the 16K ZX81	£2.50
BP115	The Pre-Computer Book	£1.95
BP116	Electronic Toys Games & Puzzles	£2.25
BP117	Practical Electronic Building Blocks – Book 1	£1.95
BP118	Practical Electronic Building Blocks – Book 2	£1.95
BP119	The Art of Programming the ZX Spectrum	£2.50
BP120	Audio Amplifier Fault-Finding Chart	£0.65
BP121	How to Design and Make your Own P.C.B.s	£1.95
BP122	Audio Amplifier Construction	£2.25
BP123	A Practical Introduction to Microprocessors	£2.25
BP124	Easy Add-on Projects for Spectrum ZX81 & Ace	£2.75
BP125	25 Simple Amateur Band Aerials	£1.95
BP126	BASIC & PASCAL in Parallel	£1.50
BP127	How to Design Electronic Projects	£2.25
BP128	20 Programs for the ZX Spectrum & 16K ZX81	£1.95
BP129	An Introduction to Programming the ORIC-1	£1.95
BP130	Micro Interfacing Circuits – Book 1	£2.25
BP131	Micro Interfacing Circuits – Book 2	£2.25
BP132	25 Simple Shortwave Broadcast Band Aerials	£1.95
BP133	An Introduction to Programming the Dragon 32	£1.95
BP134	Easy Add-on Projects for Commodore 64 & Vic-20	£2.50
BP135	The Secrets of the Commodore 64	£2.50
BP136	25 Simple Indoor and Window Aerials	£1.95
BP137	BASIC & FORTRAN in Parallel	£1.95
BP138	BASIC & FORTH in Parallel	£1.95
BP139	An Introduction to Programming the BBC Model B Micro	£2.50
BP140	Digital IC Equivalents and Pin Connections	£3.95
BP141	Linear IC Equivalents and Pin Connections	£3.95
BP142	An Introduction to Programming the Acorn Electron	£1.95
BP143	An Introduction to Programming the Atari 600XL and 800XL	£2.50
BP144	Further Practical Electronics Calculations and Formulae	£3.75
BP145	25 Simple Tropical and M.W. Band Aerials	£1.95